LEAD-BASED PAINT HAZARDS

LEAD-BASED PAINT HAZARDS

Assessment and Management

ATC Environmental Inc.

Edited by

Vincent M. Coluccio

VNR VAN NOSTRAND REINHOLD
New York

Library of Congress Catalog Card Number 94-2273

ISBN 0–442–01715–4

The material presented in this book is for informational and educational purposes only, and is not intended for use by the reader in lead hazard assessment and abatement activities. ATC Environmental Inc., the contributing authors, and their affiliates and subsidiaries have no responsibility for injuries or other damage to persons and things as a result of such activities undertaken in reliance on information provided in this book. All lead assessment and abatement projects should be professionally supervised in conformance with local, state, and federal regulations and guidelines. Neither the editor nor the contributing authors are responsible in any way for any accidental poisonings that may be claimed to have occurred as a result of recommendations made in this book.

I(T)P Van Nostrand Reinhold is an International Thomson Publishing company. ITP logo is a trademark under license.

Printed in the United States of America.

Van Nostrand Reinhold
115 Fifth Avenue
New York, NY 10003

International Thomson Publishing GmbH
Königswinterer Str. 418
53227 Bonn
Germany

International Thomson Publishing
Berkshire House, 168–173
High Holborn, London WC1V7AA
England

International Thomson Publishing Asia
221 Henderson Bldg. #05–10
Singapore 0315

Thomas Nelson Australia
102 Dodds Street
South Melbourne 3205
Victoria, Australia

International Thomson Publishing Japan
Kyowa Building, 3F
2–2–1 Hirakawacko
Chiyoda-ku, Tokyo 102
Japan

Nelson Canada
1120 Birchmount Road
Scarborough, Ontario
M1K 5G4, Canada

16 15 14 13 12 11 10 9 8 7 6 5 4 3 2 1

Library of Congress Cataloging-in Publication Data
ATC Environmental Inc.
Lead-based paint hazards: assessment and management/ATC Environmental Inc.: edited by Vincent M. Coluccio.
p. cm.
Includes bibliographical references and index.
ISBN 0-442-01715-4
1. Lead abatement. 2. Lead based paint—Removal. 3. Hazardous substances—Risk assessment. I. Coluccio, Vincent M. II. Title
TD196.L4A83 1994
828.5'2—dc20 94-2273

The opinion of this mischievous effect from lead, is at least above sixty years old; and you will observe . . . how long a useful truth may be known . . . before it is generally receiv'd and practis'd on.

Letter from Benjamin Franklin
to B. Vaughn, Philadelphia,
July 31, 1786

I . . . advance a very strong plea for painted walls and railings as the source of lead, and for the biting of fingernails or sucking of fingers, as . . . the means of conveyance of the lead to the patient.

J. L. Gibson
Australasian Medical Gazette
23:149–153 (1904)

Contents

Preface

A 1988 report to Congress, entitled "The Nature and Extent of Lead Poisoning in Children in the United States" and prepared by the Federal Agency for Toxic Substances and Disease Registry, contained the following statement: "The lead problem in the United States can be expressed in a simple statement: Lead is potentially toxic wherever it is found, and it is found everywhere."

Lead poisoning's toll on the health and productivity of our nation's population is now measured in billions of dollars. Over the next decade and beyond, a great many dwelling units and industrial structures nationwide will undergo abatement of the major source of lead affecting our nation's children—lead-based paint. The vast majority of workers who abate (remove or cover) lead-based paint or encounter lead-based paint in the course of conventional renovation work are untrained in safe techniques for handling lead and are unprotected from its hazards. All too often their work generates hazardous levels of lead particles to which the workforce and community residents are exposed.

The overall goal of this book is to impart practical knowledge of how to identify and manage lead-based paint hazards, and how to prevent further exposures to lead among the workforce and public where renovation of lead-containing surfaces is undertaken. While this book focuses on the assessment and mitigation of lead-based paint hazards, other sources of lead and their relative contribution to the nation's lead poisoning problem are also described.

This text provides guidance to individuals engaged in assessing and controlling lead hazards in housing, schools, day care centers, and commercial and industrial facilities. The contributing authors are well versed in risk communication, which in this case involves translating scientific terms and concepts related to lead poisoning prevention into "user-friendly" language that is readily understandable by the lay public.

To keep reader appeal as broad as possible, established protocols for lead-based paint detection and abatement that are published in certain regulatory and guidance documents are outlined and referenced herein, but are not reproduced verbatim. This book can serve as a supplement to the established protocols and training manuals used in lead-based paint inspection and abatement courses. It is also useful as a general reference source on lead-based paint hazards for environmental practitioners and the general public.

The case studies and the basis for insights and recommendations offered here are derived from years of practical experience by ATC Environmental Inc. and the contributing authors in investigating childhood lead poisonings, and assessing and managing lead hazards in many settings nationwide.

Individuals involved with lead hazard assessment and management are strongly urged to keep abreast of the rapidly changing regulatory requirements and technological innovations that are underway. The "Information Sources" appendix of this book will prove useful in identifying the key regulatory agencies, professional organizations, and publications where up-to-date information can be obtained.

I would like to thank Mr. Morry Rubin, president of ATC Environmental Inc., for his support of this work, and commend him for his foresight and perseverance in establishing an effective, nationwide network of lead risk-management services through ATC's regional offices. I also want to express my appreciation to the contributing authors for their efforts and their willingness to share valuable insights and technical information on the matter of lead poisoning prevention. The efforts of the text reviewers, in particular Mr. John Lange of the Center for Environmental and Occupational Training Inc. in East Pittsburgh and Mr. Peter Bass of the New York Triborough Bridge and Tunnel Authority, are also acknowledged and appreciated.

Contributors

Jackson L. Anderson, Jr., C.P.C.U., is director of finance and insurance for the National Center for Lead Safe Housing, a nonprofit organization based in Columbia, MD, whose mission is to help reduce childhood lead poisoning while preserving the nation's affordable housing stock. An honors graduate of Gettysburg College (PA), Mr. Anderson also holds the Chartered Property Casualty Underwriter (CPCU) professional insurance designation. He has more than 16 years of experience in the insurance industry, most of which has been in the field of asbestos, cumulative injury, toxic tort, environmental claims, and related insurance coverage litigation. Mr. Anderson has also held executive positions in underwriting, claims, and production with a leading asbestos and lead paint abatement liability insurer. He is a regular guest lecturer on insurance for asbestos abatement, lead abatement, and underground storage tank contracting operations for environmental engineers and consultants at several eastern universities. He has also spoken on insurance at a variety of environmental seminars, conventions, and trade shows and periodically contributes articles to various industry publications.

David Chambers, B.A., is the director of ATC's Lead Risk Assessment and Litigation Support (New York Branch) Program. He directs lead and asbestos training courses for major utilities, housing and transportation organizations, and government agencies. Mr. Chambers also provides housing and governmental clients with lead paint interim control and abatement specifications, and oversees on-site monitoring of pilot lead abatement programs for worker safety and health and environmental compliance requirements.

Vincent M. Coluccio, Dr.P.H., is vice president of ATC Environmental Inc. where he is responsible for developing lead poisoning prevention service capabilities at ATC's twenty-one branches nationwide. Dr. Coluccio is cofounder and first president of the National Lead Abatement Council, chair of a Steel Structures Painting Council task group on ambient air monitoring for lead paint abatement, and is active on national and municipal lead poisoning prevention advisory committees.

He is the director of lead inspection and abatement training courses at the National Institute for Occupational Safety and Health (NIOSH) Educational Resource Center (UMDNJ-RWJ Medical School and Rutgers) in New Jersey. Prior to working at ATC he served in environmental policy and managerial positions with the New York City Department of Environmental Protection. Dr. Coluccio earned an M.S. in environmental sciences from Hunter College, City University of New York (1981), and a doctorate in public health from the Columbia University School of Public Health (1986).

Catherine B. Ganley, B.A., is the operations manager for ATC Environmental Inc.'s New York City Lead Division. She manages and supervises environmental engineering projects related to lead hazard assessment/mitigation concerns. She coordinates and prepares waste and drinking water testing protocols and wastewater treatment systems for lead contaminated water. Under the direction of Ms. Ganley, ATC's lead division also prepares specifications for removal of lead paint from steel structures and implements high volume ambient air monitoring using TSP and PMI0 pumps. In addition to her consulting work in the lead paint industry, she has experience in the coordination and management of Phase I, II, and III environmental assessments, underground storage tank removal projects, and remediation design plans, as well as in the preparation and processing of documentation related to hazardous waste management and disposal.

David Jacobs, C.I.H., is currently deputy director for the National Center for Lead Safe Housing, where he helps direct a research program. He was as an environmental research scientist at the Georgia Institute of Technology, where he was also director of the Southern Lead Training Consortium under contract to the U.S. Environmental Protection Agency (EPA). Mr. Jacobs assisted in developing model training curricula on lead-based paint for EPA and the U.S. Department of Housing and Urban Development (HUD). He helped to write a lead-based paint risk assessment protocol for the Housing Authority Risk Retention Group; the protocol was later adopted by HUD for the nation's public housing program. Mr. Jacobs has testified before Congress several times on lead poisoning issues and has conducted a number of research efforts on abatement, dust sampling, and paint testing. He has trained more than 2,200 public and Indian housing authorities in lead-based paint issues under a comprehensive agreement with HUD and EPA and has worked on surveillance of all environmental aspects of lead-based paint abatement. Mr. Jacobs earned a B.A. in political science from Antioch College (1973), a B.S. in environmental health from Oakland University (1983), and an M.S. in technology and science policy from the Georgia Institute of Technology (1988).

Nick Malino is senior vice president and chief operating officer of ATC Environmental Inc. and is responsible for operations and finance functions at ATC's twenty-one branch offices nationwide. Prior to joining ATC, Mr. Malino managed several large clinical laboratory facilities and oversaw the development of several novel testing methodologies. He earned master's degrees in biology (1981) and finance (1986) from the University of Bridgeport, Bridgeport, Connecticut.

Christopher M. Placitella, Esq., is a partner in the law firm of Willentz, Goldman & Spitzer with offices in Woodbridge, NJ, and New York City. He is cochairman of the firm's Toxic Tort Litigation Team and currently represents lead poisoned children in litigation in both New York and New Jersey.

Keith W. Pokorny, B.S., is the manager of the Lead Services Division at ATC/Dennison Environmental Services Inc. in Boston, MA, where he has served in this capacity since 1989. Mr. Pokorny oversees the operations of all lead-related projects for ATC/Dennison's New England and New York regional offices. Prior to ATC/Dennison, Mr. Pokorny was manager of environmental remediation projects for a major environmental consulting firm in Boston for two years. He is a graduate of Clarkson University with a bachelor's of science in industrial hygiene and environmental toxicology.

Lloyd Smith, Ph.D., is vice president of Corrosion Control Consultants & Labs Inc., a firm specializing in industrial protective coatings, with special emphasis on industrial lead paint removal. He received a Ph.D. in chemistry in 1970 from the State University of New York at Buffalo and has worked in corrosion and coatings for twenty years. He joined CCC&L in July 1991, having worked previously for S. G. Pinney & Associates Inc. and the Federal Highway Administration. Dr. Smith is an active member of technical societies involved with corrosion protection and coatings, such as the Steel Structures Painting Council (SSPC), National Association of Corrosion Engineers (NACE), and American Society for Testing and Materials (ASTM). He served on the initial board of governors of SSPC and chairs the SSPC Lead Removal Committee. He is one of the primary instructors for the SSPC Lead Paint Removal from Steel Structures Tutorial and has served as principal investigator for research studies on lead paint removal from bridges and other structures.

Barry R. Sugarman, Esq., is an associate with Willentz, Goldman & Spitzer. He concentrates his practice in the areas of childhood lead poisoning and asbestos litigation. In addition to his civil cases, Mr. Sugarman is actively prosecuting a lead poisoning class action suit for injunctive relief against the City of Newark, NJ.

Maureen A. Sullivan is a loss control representative with the Fidelity Environmental Insurance Company. She worked in the loss control department as a special services administrator prior to her activity as a field representative. Ms. Sullivan served as first secretary/treasurer for the National Lead Abatement Council (NLAC) and is chairperson of its membership committee. Ms. Sullivan is also on the Lead Task Force for the NAC/The Environmental Information Association.

Mark P. Vink, Ph.D., is the vice president of loss control for the Fidelity Environmental Insurance Company. He has more than fifteen years of environmental safety experience, with assignments in government, industry, and insurance. He received a doctorate from Temple University, and presently serves as the chairman of the Environmental Information Association's Safety and Health Committee. Dr. Vink is the past president of the New Jersey Chapter of the National Asbestos Council and the South Jersey Industrial Safety Council.

Kenneth T. White, C.I.H., M.B.A., C.S.P., is the senior environmental's system analyst for Information Technology Solutions Inc. of Hampton, VA. He has been laboratory director for three environmental/industrial hygiene laboratories and senior scientist for seven companies, including a basic-research firm, a major insurance company, and a consulting firm. Mr. White is president of the National Lead Abatement Council and the former chair of its training and qualifications committee. He has served with various American Industrial Hygiene Association committees dealing with lead, including environmental lead laboratory accreditation, and chairs the ad hoc committee on sampling of environmental lead. Mr. White is a certified industrial hygienist (chemical aspects) and a certified safety professional (comprehensive practice). He has earned degrees in chemistry from Thiel College, Greenville, PA (B.A.) and the University of Rhode Island, Kingston, RI (M.S.), and a multiple-major M.B.A. from Northwestern University, Chicago, IL (management/policy, organization behavior, and marketing).

ATC Environmental Inc. is a publicly held, full service environmental management firm with twenty-one regional offices. ATC provides lead hazard assessment and strategic management programs for governmental and corporate sector clients; public policy development on lead poisoning prevention; lead litigation support; lead testing and abatement project management for nearly 100 public housing authorities; and lead training programs.

Introduction

Vincent M. Coluccio, Dr.P.H.

Lead is a metal that has been used as an additive to paint for thousands of years. Archaeologists have found buildings dating back to 3000 B.C. with lead pigments that have retained their color for 5,000 years. Accompanying this historical use of lead-based paint, however, is a recognition, also dating back thousands of years, of human health hazards associated with incidental exposures to the lead.

During most of this century in the United States, lead-based paint has been the coating of choice for residential properties. An estimated 3 million tons of lead[1] is present in paint applied to 57 million homes nationwide.[2] Lead primers are also the coating of choice for steel structures, estimated to cover at least 185,000 steel bridges,[3] more than 51,000 steel water tanks operated by community drinking water purveyors, and more than 209,000 petroleum tanks[4] (these estimates do not include industrial water tanks, which would most likely exceed 100,000). As the nation's housing stock and steel structures age, so too does the lead-based paint, releasing lead in the form of chips and dust. The cumulative exposures to lead from paints and other sources has resulted in an epidemic of worker and childhood lead poisoning across the nation.

In response to the childhood lead poisoning problem, municipalities have developed childhood lead poisoning prevention programs. Baltimore was the first city to implement blood lead screening in 1935. In 1951 Baltimore prohibited the sale of leaded paint for the interior of dwellings and, in 1958, required warning labels on cans of leaded paint. During the 1950s and 1960s other cities in the United States began to regulate the use of lead-based paint, educate the public on lead poisoning prevention, and implement blood lead screening programs. Some other cities with early regulations banning the use of lead-based paint were Chicago, IL; Cincinnati, OH; Jersey City, NJ; New Haven, CT; New York, NY; Philadelphia, PA; St. Louis, MO; Washington, DC; and Wilmington, DE.[5] During this period only acute childhood lead poisoning symptoms (such as kidney failure, convulsions, paralysis, and comas) were commonly recognized, and the cause of the lead exposure was largely attributed to ingestion of chips of lead-based paint.

The major federal initiative to combat lead poisoning began in 1971 when President Richard Nixon signed the Lead Based-Paint Poisoning Prevention Act (LBPPPA), which established a broad agenda for federal actions to combat lead poisoning. Concerted efforts to educate the public, screen millions of children for lead poisoning, and remove lead from gasoline and other consumer products have proven successful over the past three decades in markedly reducing the number of mortalities and acute childhood lead poisonings resulting from exposures to large doses of lead.

While severe poisoning of children was the typical form of the disease at one time, the lead hazard is now viewed as a more pervasive problem of lower lead exposures that were once deemed to be harmless—and are now recognized as detrimental. It must be remembered, however, that unacceptable numbers of severely poisoned children continue to be identified and treated in many clinics and hospitals of the nation's urban areas.[6]

Furthermore, our nation's lead poisoning prevention efforts have largely been "reactive" to the lead paint poisoning problem. Environmental interventions such as abating dwellings of lead paint are typically triggered only after a child has been found to have elevated blood lead, rather than based on the more cost-effective and "proactive" approach aimed at preventing lead exposures and lead poisoning. Lecturers on lead poisoning prevention often draw an analogy between this reactive approach and the use of canaries for detecting poisonous gases in underground mines, illustrating that lead poisoned children serve as "detectors" of hazardous lead conditions in their environment. Stated differently, the poisoning of the child is often a prerequisite to the identification and attempted abatement of hazardous lead conditions. Compounding this problem is the increased exposures to lead among children and adults when abatements are done improperly.

Today an estimated 3.8 million children in the United States have blood lead levels high enough to cause adverse health effects. The major source of this lead is residential lead-based paint.[7] Lead is a potent toxin that has no known beneficial function in the human body. Lead also accumulates in bones where it can remain stored for decades causing long-term health problems. Childhood lead poisoning produces lower intelligence, behavioral disorders, hyperactivity, aggressiveness, and mental retardation. It slows development, impairs hearing, and causes kidney and immune system disorders. High-dose exposures can cause convulsions, coma, and even death.

Lead poisoning is not readily treatable and in many cases the adverse health effects can be irreversible. Blood lead levels can be lowered through a form of medical treatment known as chelation therapy, but the treatment is less effective at removing lead that has already been stored in other body tissues.

This scenario begs a central question: Why has our nation waited until the 1990s to embark on an aggressive program of addressing the lead paint poisoning problem? The answer to this question is at least partly addressed by the following considerations:

- Biomedical research recently (during the 1980s) confirmed the toxicity of lead at low exposure levels, prompting federal authorities to lower the blood lead criteria for lead poisoning; this action caused a marked rise in the number of children defined as having "elevated blood lead levels."
- The long-held perception of lead poisoning as strictly an acute poisoning event associated with ingestion of paint chips has been modified; a child's incidental exposure to lead dust is now recognized as sufficient to result in more subtle—but still serious—health impairment.
- Although children in impoverished, inner-city areas are at greatest risk of lead poisoning, it is now understood that lead poisoning also affects children in all socioeconomic settings, from the poorest to the wealthiest.
- An increasing number of litigants are successful in their claims of personal or property damages due to lead contamination.
- Public awareness of the hazards of lead exposure has been heightened by extensive media coverage and government-sponsored public awareness programs.

The early 1990s have seen a renewed federal response to the demand for proactive lead poisoning prevention programs. Most notably, the U.S. Department of Housing and Urban Development (HUD) released its *Lead-Based Paint: Interim Guidelines for Hazard Identification and Abatement in Public and Indian Housing* in 1990, which provides detailed guidelines for implementing a federal strategy of abating public housing authorities of lead-based paint. In 1992 Congress altered this approach to addressing lead hazards by passing Title X (10) of the Housing and Community Development Act of 1992 (P.L. 102–550). Title X sets forth a comprehensive strategy for evaluating lead-based paint hazards through new risk assessment procedures and establishes a phased approach to controlling lead-based paint hazards through "interim control" measures to manage lead hazards over the short-term, and where feasible "abatement" to permanently eliminate the hazard.

Many states and localities are also updating their outdated and inadequate lead poisoning prevention programs, which are largely modeled on earlier federal initiatives. However, these programmatic revisions face formidable challenges by budgetary constraints, concern for the impact of lead liabilities on affordable housing, and inadequate lead-related insurance provisions for property owners and abatement firms.

While the so-called full-removal approach represents a permanent solution and is the ultimate goal in addressing lead-based paint hazards, it suffers some major drawbacks: Doing it properly is prohibitively expensive in many applications, and when it is done improperly it increases adult and childhood exposure to lead. The more affordable and conventional scrape-and-sweep approach to correcting deteriorated lead-based paint (while leaving nondeteriorated lead paint in place) also increases adult and childhood lead exposures and does not address ongoing deterioration of adjacent lead-painted surfaces. Lead hazard risk assessments coupled with appropriate hazard controls offer the most practical and promising means of managing lead-based paint hazards where full abatement is not practicable.

The risk assessment procedure provides information and advice to the owner of a dwelling unit so that quick action can be taken to control lead hazards posing an immediate danger to occupants or building maintenance workers. In order to provide the necessary guidance, a risk assessment is composed of three parts:

1. Identification of the existence, nature, severity, source, and location of immediate lead-based paint hazards
2. Identification of the owner's management and maintenance practices that may require modification to address lead-based paint
3. A determination of the various options for feasible hazard control, including interim control and abatement measures, if hazards are identified.

The risk assessments may point to full abatement or removals in some units, and interim controls in other units. *Interim controls* refer to methods that safely cover or render a leaded surface intact, followed up with maintenance and surveillance of the surface until such time the surface will be permanently abated. Proper interim control programs afford far greater protection to workers and occupants than do conventional scrape-and-repaint methods. Risk assessment and interim control programs may represent the only realistic and affordable option to addressing lead-based paint in much of our nation's housing.

While the major impetus for government action in the 1990s is the protection of children from lead hazards, our discussion on lead poisoning is not restricted to children. Many adult cases of occupation-related lead poisonings are reported; many more cases go unreported. Worse yet, worker protections from lead hazards in both the construction and general industry settings have gone largely underenforced.

Some recent progress at the federal level can also be reported for worker protection. In 1993 OSHA took the long-awaited step of strengthening the lead standard for construction workers by promulgating the "Lead in Construction; Final Interim Rule."[9] These standards are comparable to those specified in the general industry standard and the HUD guidelines for lead paint abatement. A commitment at the federal, state, and local levels to implement these new standards will afford much needed health protection to 900,000 construction workers nationwide.

The proper disposal of lead abatement wastes is a matter of great concern, confusion, and liability to building owners and lead abatement contractors. Classifying abatement waste as hazardous waste because of its lead content can in some cases boost disposal costs fivefold or tenfold compared to disposing of it as construction debris or nonhazardous solid waste. Further scientific research and a national concensus on rational and affordable disposal strategies are required; the discussions and case studies presented in this text provide some practical perspectives on the disposal dilemma.

Notes

1. Agency for Toxic Substance and Disease Registry. *The Nature and Extent of Lead Poisoning in the United States: A Report to Congress.* July 1988. p. II-5.
2. U.S. Department of Housing and Urban Development. *Comprehensive and Workable Plan for the Abatement of Lead-Based Paint in Privately Owned Housing. A Report to Congress.* December 7, 1990. p. xvii–xix.
3. E. D. Carlson. Testimony before the Subcommittee on Health and the Environment, Committee on Energy and Commerce, U.S. House of Representatives, Washington, DC, March 1991.
4. CONSAD Research Corporation. *Economic Analysis of OSHA's Interim Final Standard for Lead in Construction,* final report for contract number J-9-F-1-0011. OSHA, April 1993.
5. J. F. Gilsinn. *Estimates of the Nature and Extent of Lead Poisoning in the United States.* U.S. Department of Commerce, National Bureau of Standards, 1972, table 1, p. 11. (Cited in: U.S. Department of Housing and Urban Development. *Comprehensive and Workable Plan for the Abatement of Lead-Based Paint in Privately Owned Housing. A Report to Congress.* December 7, 1990. p. 1–2.)
6. Agency for Toxic Substance and Disease Registry, p. 43.
7. Centers for Disease Control. *Preventing Lead Poisoning in Young Children.* October 1991, p. 12.
8. *Federal Register,* May 4, 1993. 29 *CFR* 1926.62.

1

Sources of Lead in the Environment

Vincent M. Coluccio, Dr.P.H.

Lead, a soft, bluish-gray naturally occurring heavy metal, is considered to be one of the seven metals of antiquity. The chemical symbol for lead as shown on the periodic table of the elements is Pb, from the Latin word for the metal, *plumbum*. The atomic weight of lead is 207.

The predominant lead ore is a combination of lead and sulfur called galena. Other lead-containing minerals include anglesite ($PbSO_4$); cerussite, also known as lead carbonate or "white lead" ($PbCO_3$); mimetite [$PbCl_2 \cdot 3Pb_3$-$AsO_4)_2$]; and pyromorphite [$PbCl_2 \cdot 3Pb_3(PO_4)_2$]. The four isotopes of lead are Pb 204, 206, 207, and 208.[1]

The addition of lead to paint and many other consumer products such as gasoline and pesticides has improved the overall quality of those products in spite of the known or anticipated health effects arising from incidental exposures to the lead. This chapter reviews the major uses of lead and the means by which the environment becomes contaminated with lead.

LEAD AS A PAINT ADDITIVE

The major constituents of paint include vehicles, solvents, pigments, and additives.[2] *Vehicles* contain a combination of resins, plasticizers, drying oils, and so on, that provides film continuity and adhesion to the substrate. They are called vehicles because they carry to the substrate the ingredients that will remain on the surface after the paint has dried. *Solvents* are added to enhance paint application properties. *Pigments* provide a decorative function by contributing opacity, color, and gloss, and a protective function by contributing hardness, resistance to corrosion, resistance to abrasion and weathering, and improved adhesion. *Additives* are ingredients that impart desirable qualities to the wet paint or dried paint film by modifying the properties of either the vehicle or the pigmentation, or both. For example, additives may increase the speed of drying, increase resistance to fading, or facilitate application of the wet paint.

Lead-based paint has been the preferred coating medium in millions of homes. Addition of lead to paint imparts many desirable characteristics to the paint film. Leaded paints adhere well to a variety of surfaces and may last longer than paints without lead. Lead pigments produce a bright and durable paint, and by virtue of its toxicity, lead forms an effective barrier against the growth of mildew. Perhaps the best demonstration of these qualities of lead is a finding by archaeologists that lead-based paints applied on buildings constructed 5,000 years ago still retain recognizable colors.[3]

Lead acetate is added as a drier and extender. In varnishes it can extend the surface coverage by as much as 30 percent. Lead acetate is a sweet tasting compound and has been referred to as "sugar of lead" and its presence in paint chips encourages children's ingestion of the chips.

Lead carbonate (cerussite) was commonly used as a pigment in residential paints in the United States until 1977. Lead sulphate (also known as *sublimed white lead*) was also added to paint and it chalks less than lead carbonate.

Lead-containing paints are also the primary choice of protection against metal corrosion. Red lead is an orange-red colored mixture of the toxic elements lead and chromate, used extensively as a primer for metal surfaces. Omaha Orange paint also has a very high lead content and is used on school buses.

Release of Lead-Based Paint Dust and Chips

Lead that is bound within an intact film of lead-based paint represents a potential source of lead particles for human exposures. The actual hazard manifests itself when individuals are exposed to lead particles that are released from the film. Lead particles are released from paint in the following ways:

- When used on exterior surfaces, certain lead-based paints tend to chalk, that is, their outer layers gradually slough off thereby revealing a fresher and better-appearing outer layer of paint. Ultraviolet light and moisture enhance chalking by causing a microscopic layer of paint to separate from its polymer binding. While this process provides a desired aesthetic outcome, the lead dusts can accumulate to problematic concentrations in adjacent soils.
- Peeling and flaking of paint increases the surface area for chalking to occur and can result in release of paint chips. Lead-based paint on friction surfaces such as door and window components and floors, and on surfaces damaged by moisture, vibration, or impact, can be expected to gradually release lead particles.
- Disturbance of lead-based paint in the course of conventional renovation and

demolition activities can generate large volumes of lead dust particles and chips, as can improper abatement practices. The removal of lead-based paints from steel structures is often accomplished by abrasive blasting, which in turn can cause high levels of air and soil contamination.

Once lead-based paint has been transformed to dust, the normal actions of a small child put it at great risk of exposure. The normal hand-to-mouth activity displayed by young children increases the chance of lead-dust ingestion. This, coupled with crawling on the floor and a child's tendency to place its mouth on windowsills, door frames, and the like, may cause heavy exposures to lead. Children who suffer a psychiatric disorder called "pica" have a tendency to ingest nonfood items and are at especially high risk of ingesting lead-containing materials.

Research findings on the contribution of lead-based paint to childhood lead poisoning as reported in a federal report to Congress[4] are summarized as follows:

- Eating chips of lead-based paint can result in severe poisoning; however, such episodes are relatively infrequent.
- Ingestion of dust and soil containing lead through hand-to-mouth activity is a more common pathway among children than eating paint chips.
- Researchers have found significant associations between lead in children's blood and lead on their hands, and in the dust and soil in around homes.
- There is evidence that homes in poor condition elevate the hazards of exposure to lead-based paint.
- Home finishing (scraping and repainting), if not properly performed, can significantly increase the hazard level (see Fig. 1-1).

While different research studies and regulatory strategies target different age groups of children, usually "under 7" or "under 6" years of age, all children including those older than 7 years are susceptible to lead exposure and poisoning.

Extent of Lead-Based Paint Use

Since 1910 an estimated 4.2 million tons of lead were used for white paint alone in the United States and about 3 million tons of lead probably remain in paint in our nation's housing.[5] The results of a federally sponsored survey[6] on the extent of lead-based paint use and associated lead contamination are summarized as follows:

- Of the 77 million privately owned and occupied homes built before 1980, 57 million, or three-fourths, contain lead-based paint. Of these 57 million units, an estimated 9.9 million are occupied by families with children under the age of 7, who are at most risk of lead poisoning.
- Of the units containing lead-based paint and occupied by young children, 3.8 million have peeling paint, excessive amounts of dust containing lead, or both.

NATIONAL LEAD INFORMATION CENTER

HOME REPAIRS AND RENOVATIONS: WHAT YOU SHOULD KNOW ABOUT LEAD-BASED PAINT

Any house or apartment built before 1978 may contain lead-based paint. **Most** homes built before 1960 contain lead-based paint. Lead-based paint produced before 1960 contains higher concentrations of lead than paint manufactured in later years.

Lead-based paint can be on walls, ceilings, woodwork, windows, and sometimes floors. When lead-based paint on these surfaces is broken, sanded, or scraped, it breaks into tiny, sometimes invisible, pieces that your child may swallow or inhale. Even small repair and renovation jobs, including repainting projects, can create enough lead dust and chips to harm your child.

BEFORE YOU REPAIR OR RENOVATE

BEFORE you disturb a surface with old paint on it, you should, if possible,

- Call your local health department and ask if they can test your home for lead-based paint. If the health department cannot test, ask them who can.
- If lead-based paint is found in your home, have the repair or renovation done by a worker who has been trained to protect your family and home from exposure to lead dust and chips.

IF YOU FIND OR SUSPECT THAT LEAD-BASED PAINT IS PRESENT

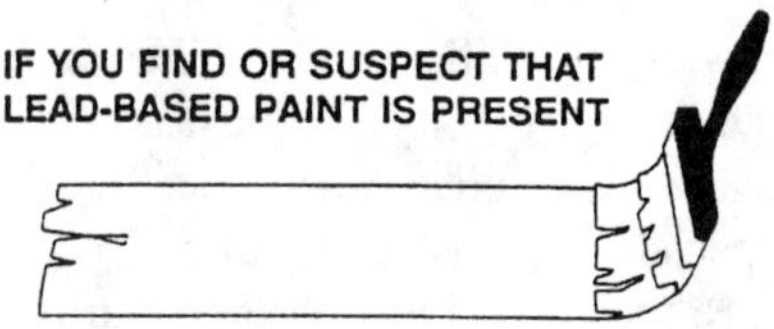

You should AVOID the following activities, which can produce paint dust and chips, in areas of your home where you know or suspect there is lead-based paint:

- scraping, sanding, or using a heat gun on painted surfaces before repainting;
- making holes in walls to get at pipes or install electrical outlets;
- tearing out walls;
- repeatedly bumping furniture or other objects into painted walls;
- unnecessarily opening and closing windows with painted frames and sills.

If you <u>must</u> do repairs or renovations yourself in areas where you know or suspect lead-based paint is present, you SHOULD

- Move children and pregnant women to another apartment or house until work is completed and the area is properly cleaned.
- Cover exposed areas. If the area is small, such as an electrical outlet, keep that area covered until repair and cleanup are completed. If the area being worked on is large, such as a wall being torn down, use plastic coverings to seal off entrances and ducts and to protect furniture, carpets, rugs, and floors from paint dust and chips. Dispose of the plastic carefully.

The National Lead Information Center is operated by the National Safety Council with funding from the Environmental Protection Agency, the Centers for Disease Control, the Department of Housing and Urban Development, and the Department of Defense.

FIGURE 1-1. National Lead Information Center fact sheet concerning lead-based paint in homes.

- To keep dust down, wet painted surfaces before you work on them.
- Clean up thoroughly.

  Always clean up dust and chips with wet mops or rags soaked in a solution of trisodium phosphate (TSP) or phosphate-containing powdered dishwasher detergent and warm water. (Powdered dishwasher detergents are recommended because most have high phosphate contents. Most multipurpose household cleaners are not effective in cleaning up lead dust.) To avoid skin irritation when cleaning up with TSP or high-phosphate dishwasher detergent, wear protective gloves.

  Use two buckets--one for wash water and one for rinse water. Always wring dirty water into the wash water bucket.

  To prevent recontamination of cleaned surfaces, wash mops and rags thoroughly after each use. If this is not possible, or if you have already used the mops and rags several times, place them in plastic bags and dispose of them carefully.

- Avoid sweeping or vacuuming the work area. Sweeping spreads lead dust around. Vacuuming also spreads lead dust around, since tiny lead particles can pass through and out of ordinary vacuum cleaners.

IF REPAIRS OR RENOVATIONS HAVE ALREADY OCCURRED

If repairs or renovations of areas you know or suspect contain lead-based paint have already occurred or are occurring in or around your home, you should do the following:

- Keep children away from paint dust and chips.
- Clean up all dust and chips with wet mops and rags, as described above. Pay special attention to floors and to window sills and window wells (where the bottom of the window sash rests when the window is closed).
- Close your windows if work is going on outside your home that may be scattering lead dust--for example, a neighbor scraping exterior paint. Using wet mops and rags, clean up any dust that has gotten into your home.
- Have your children under six years old tested for lead. To arrange for testing, call your doctor or your local health department.

10/92

- There is no correlation between the incidence of lead-based paint and the income of the household. Lead-based paint is found as often in the homes of the wealthy as the poor.
- More units have lead-based paint on the exterior than the interior. Of the 57 million units with lead-based paint, 18 million have it only on exterior surfaces, 11 million only on interior surfaces, and 28 million on both exterior and interior surfaces.
- Older homes are more likely to have lead-based paint than newer homes. An estimated 90 percent of dwelling units built before 1940 have lead-based paint in the interior or on the exterior, while 62 percent of homes built between 1960 and 1979 have lead-based paint.
- Lead-based paint is found less often in the southern United States than in other regions, because the South was significantly less urban than other regions of the country until the 1970s and its housing stock on the average was built more recently.
- Lead-based paint appears most often on windows (interior and exterior), soffit and fascia, and exterior walls.
- An association exists between the presence of lead-based paint and the presence of excessive levels of lead in dust and soil. Approximately 14 percent of all housing units built prior to 1980, or 10.7 million homes, have lead in the interior surface dust that exceeds federally established guidelines for allowable levels of lead dust following residential abatements.
- The chance of a home having excessive lead dust is about twice as large if the home has high levels of interior lead-based paint than if it does not. However, most of the homes with interior dust have it only on the window sills or in the window wells. About 1 million units have excessive lead dust exclusively on the floors.
- Approximately 16 percent of all homes built prior to 1980 have concentrations of lead in soil adjacent to the house that exceed U.S. Environmental Protection Agency guidelines. The chance of this occurring is at least four to five times greater if the house has exterior lead-based paint than if it does not.

OTHER SOURCES OF LEAD

Lead is widespread in the environment as a consumer product and in our society's air, water, and solid waste streams. Once introduced into the environment, lead retains its toxic qualities and does not transform into a harmless chemical. In addition to being well-suited as an additive to paint, lead is a durable and malleable metal that is very easy to work with. Lead readily mixes with other metals to form alloys and it does not rust. Lead pipes expand when water within them freezes, thereby preventing bursting.

Gathering information about the major sources of lead contamination other than lead-based paint is essential to the conduct of childhood lead poisoning investigations considering the myriad sources of lead exposures in addition to paint. This information is also important in understanding how environmental sources of lead contamination other than lead-based paint can influence lead dust levels on abatement sites, at times making it difficult or impossible to achieve postabatement cleanup standards.

Occupational Exposure

Some affected occupational categories include:

Battery manufacturing and repair
Demolition/renovation of old homes or other structures
Automobile assembly, auto body and radiator repair
Secondary lead smelting (recovery of lead from batteries)
Lead scrap melting or metal founding
Inorganic chemical industries
Ammunitions manufacturing
Brick making
Cable making and splicing
Cutlery manufacturing
Fish sinker manufacturing
Jewelry making
Plastic manufacturing
Pottery making
Roofing
Glass and stained glass manufacturing
Welding
Law enforcement
Plumbing
Rubber manufacturing
Ship building

Lead Plumbing

Lead in drinking water typically originates in building service connections (i.e., the pipe connecting the building plumbing system to a water main) and plumbing fixtures, not in reservoirs, aquifers, or other sources of water. Millions of homes have copper piping containing lead solder. Brass and bronze are alloys of metals including lead, and lead can leach out of fixtures made of these alloys. In addition, some older water coolers and drinking fountains were made with lead fixtures or lead-lined tanks.

The federal Safe Drinking Water Act of 1986 prohibited the use of lead piping and lead solder in drinking water plumbing systems. However, much lead still remains from past uses, and despite the lead ban it can be expected that old inventories of lead products will continue to be added, either through ignorance or fraud, to plumbing systems.[7]

The EPA has ordered a recall of lead containing coolers and certain other drinking water devices and has established 15 parts per billion of lead in drinking water as an "action level" for water suppliers to begin implementing measures to reduce consumer exposure to lead in drinking water.[8] EPA requires that water supply agencies or companies frequently test their water for lead content and many other parameters and has developed protocols for testing drinking water systems in schools for lead content. Lead hazard assessors will find useful information about drinking water lead concentrations for a particular area by contacting the water supplier and perhaps area schools that have performed water testing. In addition, EPA has a Safe Drinking Water Hotline phone number: (800) 426–4791.

A major source of lead in drinking water is the solder used in plumbing fixtures. The amount of lead in the solder may easily reach or exceed 50 percent. The lead will slowly leach into the water system from the solder, the rate of leaching depending on factors such as water pH (the more acidic the water the greater the leaching), water corrosivity, water retention time in the plumbing systems, and so on. High levels of lead may be found in the first "flush" of water from the tap, but this usually dissipates quickly as the water is allowed to run.

Lead in drinking water is thought to contribute on the average about 20 percent of the lead body burden among children. This contribution from drinking water is projected to decrease substantially as EPAs initiatives are implemented. In rare instances lead in drinking water can cause severe lead poisoning; one possible scenario is where parents routinely prepare baby formula with "first flush" water that has been allowed to sit for extended periods of time in high content lead plumbing.

Leaded Gasoline

Lead was added to gasoline as an antiknock compound beginning in the early 1920s. Leaded gasoline was banned from use in post-1975 vehicles (although pre-1975 vehicles and farm vehicles may still use it) with the advent of catalytic converters. The reduction of lead in gasoline has significantly lowered airborne lead concentrations and consequently the average blood lead levels in the United States.

Although use of leaded gasoline is largely curtailed, the decades of leaded gasoline emissions have taken their toll on the environment. Soils along heavy traffic routes have very high lead levels in spite of the ban nearly 20 years ago. An EPA study released in 1988 provided an estimate of 4 to 5 million metric tons of lead from automotive exhausts present in the nation's dust, soil, and sediments.[9]

Food Contamination

Lead from various sources may be present in foods. The solder on aluminum cans imported from foreign countries where lead use is not restricted may contain lead. Lead arsenate was used as a pesticide in the past on certain crops and the lead still remains in soil; fruits and vegetables grown in lead-contaminated soil may take the lead from the soil as they grow. Plant uptake of lead is a complex function of soil type and soil pH, plant type, chemical composition of the lead compounds, and other factors described elsewhere.[10]

The glazes used on pottery and ceramics may contain lead. Of particular concern is the glaze used on products from other countries that may not have as strict controls as the United States. Lead crystal glasses and decanters often contain very high levels of lead that may leach into the liquid poured into the container.

Toys

Some toys have been painted with lead-based paint in the past and, as with pottery or ceramics, a particular concern is with toys from another country. They can and do slip into the United States in spite of testing standards. Antique toys are also a concern, including those that were actually made of lead (soldiers, novelty banks). Finally, although a toy may not contain lead or be painted with lead-based paint, they can be easily contaminated by household dust containing lead, and a child playing with the toy is at risk of ingesting such lead.

Stationary Point Sources

Mainly smelters, stationary point sources cause localized deposition of lead particles in a community and result in elevated blood lead levels in relatively small and local populations. Exposures are due in part to current emissions and in part to resuspension of dusts and soil contaminated by past emissions.[11]

Superfund National Priority List (NPL) Sites

More than 800 of these sites have lead identified as a contaminant and may have very high soil lead levels. Information on locations and conditions of hazardous waste sites can be obtained by calling the EPA Superfund hotline number, (800) 424–9346.

Miscellaneous Sources

Lead has been used in thousands of products and industries, each of which is a potential source of lead contamination. These sources include:

Colored newsprint
Cosmetics ("Surma," a black eyeliner containing 85% lead)
Ammunition, firing ranges
Radiation shields, sheet lead for sound insulation
Power stations, battery factories
Folk remedies
Curtain weights
Colored bread wrappers
Wrapping paper
Temporary hair colorings

CONCLUSION

In many cases of lead poisoning, lead exposures may have resulted from multiple sources of lead with each source contributing a different percentage of the total exposure. Lead is so prevalent in our environment that singling out the predominant source of exposure is sometimes a difficult task. In each case, an investigator or health official must examine the activities of the individual affected and attempt to piece together all the potential sources of contamination encountered.

Notes

1. M. Windholz, ed. *The Merck Index,* 9th ed., Rahway, NJ: Merck & Co., 1976, p. 5235.
2. G. E. Weismantel, ed., *Paint Handbook,* New York: McGraw-Hill, 1981, p. 1–12.
3. University of Cincinnati, National Institute for Occupational Safety and Health Educational Resource Center. *Instructor Manual: Lead Abatement Training for Supervisors and Contractors.* July 1992, p. B-2.
4. U.S. Department of Housing and Urban Development. *Comprehensive and Workable Plan for the Abatement of Lead-Based Paint in Privately Owned Housing. A Report to Congress.* December 7, 1990, p. 2–16.
5. Agency for Toxic Substance and Disease Registry. *The Nature and Extent of Lead Poisoning in Children in the United States: A Report to Congress.* July 1988, p II-5.
6. U.S. Department of Housing and Urban Development, p. xvii–xix.
7. ATC Environmental Inc. has investigated cases where lead was applied to building water distribution systems during the 1990s.
8. *Federal Register* 56., 110, June 7, 1991, pp. 26460–26584.
9. U.S. Environmental Protection Agency (1986). *Air Quality Criteria for Lead,* EPA report no. EPA-600/8-83/028F-dF. Research Triangle Park, NC: Office of Health and Environmental Assessment, Environmental Criteria and Assessment Office, June 1986; addendum, September 1986.
10. R. Chaney, S. Sterrett, and H. Mielke. The potential for heavy metal exposure from urban gardens and soils. In *Proceedings of a Symposium on Heavy Metals in Urban Gardens*, J. R. Preer, ed. Washington, DC: University of the District of Columbia Extension Service, 1984, p. 37–84.
11. U.S. Environmental Protection Agency. *Strategy for Reducing Lead Exposures.* February 21, 1991.

2

The Health Consequences of Lead Exposure

Vincent M. Coluccio, Dr.P.H.

Nick Malino, M.S.

Lead serves no known beneficial purpose in the human body. As opposed to certain other metals that humans require in order to function (e.g., iron, zinc, copper), lead exhibits, many scientists believe, a "continuum of toxicity." In other words, the slightest exposure to lead is thought to trigger biochemical changes that may have an adverse effect on some organ of the body.

Epidemiological studies have not clearly shown effects of lead at very low levels, even though they are thought to occur and result in "subclinical toxicity" or harmful effects that are not detected on a standard clinical examination. Biochemical studies have shown inhibition of enzymes at levels lower than the current Centers for Disease Control (CDC) threshold blood lead toxicity level. It is difficult to measure and quantify health effects of lead at very low levels, epidemiologically, resulting in considerable controversy. However, health effects associated with higher lead exposures as discussed below are well characterized.

ROUTES OF EXPOSURE

There are two major routes of entry for lead into a child's body—ingestion and inhalation. Absorption of lead may occur through skin contact with an organic lead compound such as tetraethyl lead, but this type of exposure would most likely occur among adults in certain occupational settings and is not common among children.

Ingestion

The normal oral exploratory behavior of children places them at risk of lead exposure through direct ingestion of lead, as in the case of eating chips of leaded paint, or through their indirect exposure in lead-contaminated dust or soil, whereby children inadvertently transfer lead to their mouths and swallow it. Children can also ingest lead from drinking water, food, and other sources described in Chapter 1.

Some young children have a condition known as *pica* that places them at greater risk of lead exposure. Pica is a psychiatric disorder characterized by the ingestion of nonfood items, which increases the risk of a child ingesting leaded paint chips or other lead-containing items. Pica should not be confused with the normal mouthing tendencies of children.

About 50 percent of the lead ingested by children moves directly into the "circulating pool" of blood, where it passes from organ to organ; for adults, the percentage of absorbed lead drops to 8–10 percent. The nonabsorbed portion of ingested lead will be eliminated via the gastrointestinal tract. Particle size influences the degree of lead absorption by the gastrointestinal tract: Smaller particles of lead tend to be absorbed more efficiently than larger particles.

Inhalation

Some portion of airborne lead particles that are inhaled will be deposited in the respiratory tract. Studies suggest that the rate of deposition is about 30–50 percent for adults and is determined by factors such as particle size and ventilation rate[1] (little is known about the rates of deposition in children). Virtually all of the lead particles that are deposited in the lower respiratory tract (where the exchange of gases takes place) will be rapidly absorbed into the body. Lead particles that are deposited in the upper respiratory tract will be carried up the airways on mucous and either swallowed, and thereby enter the gastrointestinal tract where they may be absorbed, or expectorated.

TRANSPORT, RETENTION, AND EXCRETION OF LEAD

Once lead enters the human body it is distributed to three different components: blood, soft tissues, and bone. The "body burden" of lead refers to the total amount of lead in these three components.

Almost all of the lead in blood is associated with the erythrocytes (red blood cells).[2] The half-life of lead in blood is approximately 36 days for adults, while the half-life for a 2-year-old child may be as long as 10 months.[3]

Lead often follows the same path as calcium. Thus, the remainder of the absorbed lead is stored in the bones with a half-life of about 25 years. About 90–95 percent of an adult's body burden of lead is stored in the bones; for children the figure is about 70 percent. The bone-based lead was long thought to be inert, but the bones are living tissue and sensitive to lead toxicity. Many conditions can rapidly mobilize stored lead into the blood, including pregnancy, nursing, menopause, puberty, osteoporosis (demineralization of the bone), and even a fracture of the bone. Therefore, bones are considered a long-term reservoir of lead and can be the source of lead poisoning many years in the future.

Small amounts of lead are also stored in the teeth and in soft tissues such as the kidney, liver, and parts of the brain. Stored lead is excreted from the body by the kidneys at a very slow, steady rate and smaller amounts of lead are eliminated through the gastrointestinal tract via the bile duct. Because of this slow elimination process, individuals who are removed from lead contaminated environments due to high blood lead levels may not achieve normal levels for months, sometimes years.

MEASURING LEAD IN THE BODY

Blood Lead Levels

The most widely accepted and commonly used measure of internal lead exposure is the concentration of lead in the blood, conventionally denoted as micrograms of lead per deciliter of whole blood. A *microgram* is one millionth of a gram, or about one-tenth the weight of a grain of salt, and is abbreviated "μg." A *deciliter* is one-tenth of a liter and is abbreviated "dl." Therefore, micrograms of lead per deciliter of blood is abbreviated as "μg/dl" (see Fig. 2-1).

Blood lead levels indicate the circulating lead levels, not the stored levels in the bone and teeth. In other words, the blood lead level does not reveal lifetime history of exposure but provides a "snapshot" of the most recent exposure. However, blood lead levels will be influenced by lead that is released from bone and other storage sites in the body, so elevated blood lead levels may be measured even in the absence of recent exposures to lead.

Erythrocyte Protoporphyrin Levels

Another blood measurement defining lead toxicity is the erythrocyte protoporphyrin (EP) level, which provides information on the "total body burden" of lead. An erythrocyte is a red blood cell and protoporphyrin is a substance within the red blood cells. The addition of iron to protoporphyrin results in the formation of *heme,* a component of the hemoglobin molecule. Lead prevents the iron from combining with the protoporphyrin, thus impairing the manufacture of hemoglobin. As a result, the protoporphyrin accumulates in the erythrocyte. The EP level in the blood begins to elevate, indicating impaired heme synthesis, which may be due to lead. EP levels may be elevated by iron deficiencies and other illnesses as well, so an EP level may be elevated in the absence of excessive lead exposure.

While the blood lead level is an indicator of lead absorption, the EP level indicates a biochemical effect of the lead. However the EP test is not sensitive enough to identify children with elevated blood lead levels below about 25 μg/dl, which is

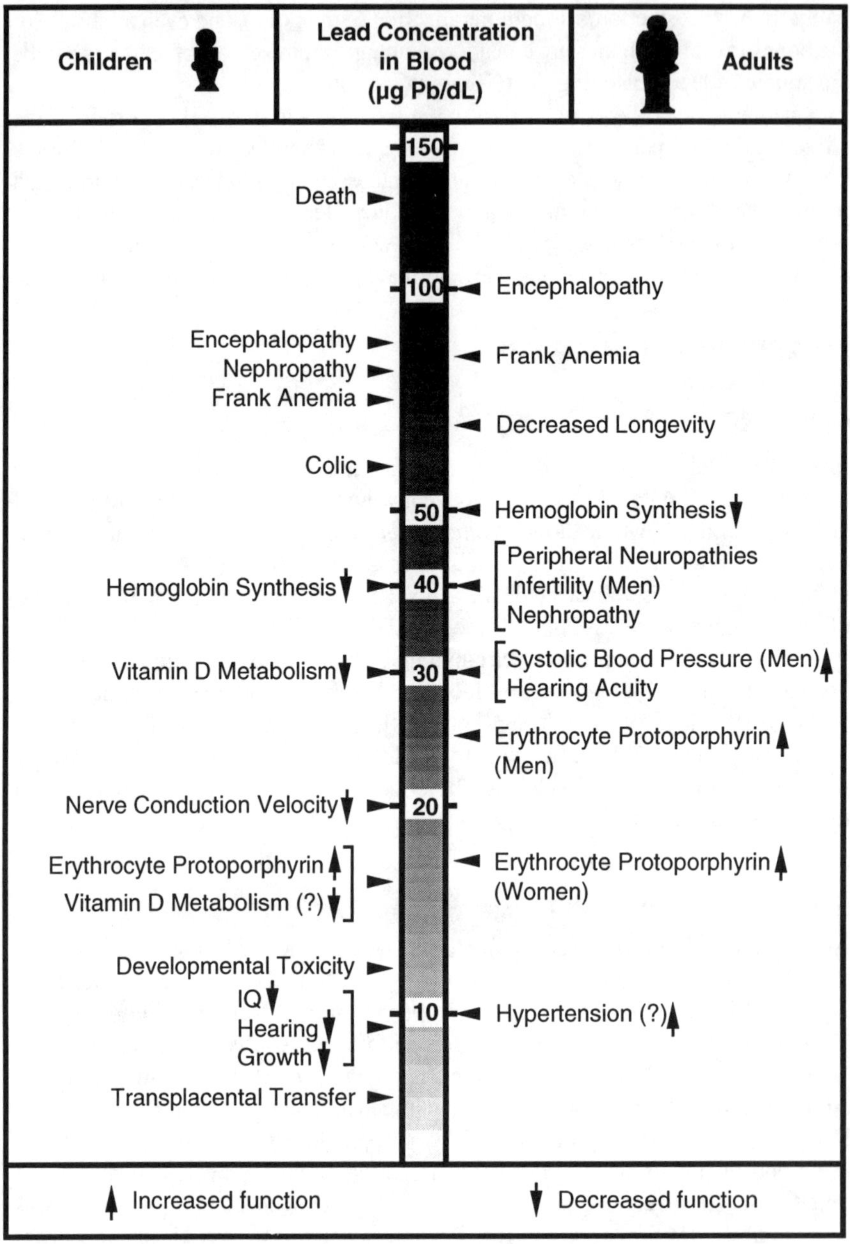

FIGURE 2-1. Potential effects of inorganic lead on children and adults: lowest observable adverse effect levels. After Agency for Toxic Substances and Disease Registry, *Toxicological Profile for Lead,* Atlanta: U.S. Department of Health and Human Services, Public Health Service, 1988.

higher than the levels considered toxic to children as discussed below, and therefore the screening test of choice for childhood screening programs is now blood lead measurements.[4]

X-Ray Testing

Another method of measuring lead in the body is X-rays. An abdominal X-ray might indicate radiopaque foreign materials, such as a lead paint chip ingested by a child in the previous 24–36 hours. Obviously the X-ray needs to be taken soon after the child ingests the lead paint chip, something that may be impossible for a parent to know. Another method of using X-rays to indicate lead exposure is an X-ray of the bones. This may reveal "lead lines" that indicate chronic and extensive exposure to lead. It is important to note that a negative result in an X-ray should not be used to rule out lead poisoning.

Lead in Teeth

Lead accumulates in children's teeth during the period of tooth eruption until the time teeth are shed. Analysis of the lead content of teeth reflects exposure to lead. For this reason teeth lead data are used as indices of cumulative lead exposures in epidemiological studies of childhood lead poisoning, but are not considered as a basis for regulatory action involving individual cases of lead poisoning.

Analytical Techniques

Current laboratory analytical techniques are available to accurately and precisely measure lead in blood and other biological media at parts-per-billion concentrations. These techniques include atomic-absorption spectrometry (AAS), anodic-stripping voltammetry (ASV), and thermal ionization mass spectrometry (TIMS). A report of the National Research Council[5] provides an in-depth assessment of techniques for quantitative measurement of blood lead levels and the biological markers of lead exposures.

TARGET ORGANS

After absorption of lead through ingestion or inhalation, lead is carried by the bloodstream to the organs of the body. Since lead is a metal it will not be destroyed during this journey and will not be altered to a nontoxic substance. Although the body has the ability to get rid of lead through the gastrointestinal system and the kidneys, in many cases an individual will absorb more lead than the body can dispose of quickly. Therefore lead is deposited in various organs and may cause irreparable damage to the organ. It is also stored in the body, increasing an individual's lifetime body burden of lead. As mentioned earlier, this chronic buildup of lead can cause health effects over time.

Once in the body, the target organs discussed below can be adversely affected by lead (see Fig. 2-2).

The Nervous System

The two major components of the human nervous system are the central nervous system, which includes the brain and spinal cord, and the peripheral nervous system, which comprises the nerves extending from the spinal chord. Neither system has any apparent barrier to or protection from lead intake.

The central nervous system is the primary target organ for lead toxicity in children. The effect of high lead concentrations on the central nervous system may include irreversible mental retardation and seizures, coma, and even death. Among children, lead exposures in the range of 80 μg/dl and above can also cause encephalopathy (brain disease), and convulsions.[6] Prior to drug therapy to remove large amounts of lead from the body, severe lead poisoning with resulting encephalopathy had a 65 percent mortality rate. This rate has significantly declined to less than 5 percent.

Epidemiological studies have shown that blood lead levels of 10–25 μg/dl can adversely affect children in a variety of ways, usually with no overt signs or symptoms of lead poisoning, including:

Subtle behavior changes such as irritability and a shortened attention span
Suppression of appetite with resulting weight loss
Sleep disturbance
Hyperactivity
Hearing impairment
Reductions in intelligence and short-term memory

Children with moderate exposures (blood lead levels of 30–50 μg/dl) are at an even greater risk of requiring special education, dropping out of school, and having reading disabilities as a young adult. Unfortunately the majority of children with elevated blood lead levels are either mildly symptomatic or asymptomatic. Exposures to lead may cause very general symptoms such as headaches or stomach aches and many health care providers do not know how to recognize or treat lead poisoning. As a result, the lead exposures can continue and exacerbate the health problem and may not be considered as the cause of observed health and/or behavioral problems.

A more common effect of lead to the nervous system occurs within the peripheral nervous system. The lead causes a condition known as ankle or wrist "drop." The wrist can no longer support the upright extension of the hand and the ankle

NATIONAL LEAD INFORMATION CENTER

QUESTIONS PARENTS ASK ABOUT LEAD POISONING

BACKGROUND

Q: **I heard that most children have less lead in their blood now than 20 years ago. If this is so, why is childhood lead poisoning in the news now?**

A: The average blood-lead level of U.S. children has come down during the last 20 years, largely because of the reduction of lead levels in gasoline and food. **But as blood-lead levels have come down, concern about the effects of low lead levels in children has risen.**

Q: **Why has concern risen about low lead levels in children?**

A: There is new evidence that lead is harmful at blood levels once thought safe. Studies show that groups of children with higher lead levels are likely to have lower IQ scores, slower development, and more attention problems than similar children with lower lead levels. These effects are subtle and have been observed in large groups of children with lead levels at least as low as 10 micrograms per deciliter. (Micrograms per deciliter, written as μg/dL, indicates the amount of lead in a deciliter of blood.)

BLOOD TESTS

Q: **Should I have my child's blood tested for lead?**

A: The only way to know for sure if your children have elevated blood-lead levels is to have them tested. The Centers for Disease Control recommend testing almost all children at 12 months of age, and, if resources allow, at 24 months. Screening should start at 6 months if the child is at risk of lead exposure (for example, if the child lives in an older home built before 1960 which has peeling or chipping paint). Decisions about further testing should be based on previous blood-lead test results and the child's risk of lead exposure. In some states, more frequent lead screening is required by law.

Q: **Why should I have my child screened for lead?**

A: Virtually all children in the United States are at risk for lead poisoning. As a result of industrialization, lead is widespread in the environment. Lead is harmful to the developing brain and nervous system of fetuses and young children. Children are more likely than adults to be exposed to lead because they have more hand-to-mouth activity than adults and because they absorb more lead than adults. Large numbers of U.S. children continue to have blood-lead levels in the toxic range. It is important for you to know if your child is one of them.

Q: **My child's blood-lead level is between 15 and 19 μg/dL. The doctor tells me to bring her back in a few months for another test. Isn't there anything else I could be doing?**

A: Your child's test showed that she had more lead in her blood than the average child in the United States. If your child's level stays in this range for several months or goes up, you should have your home inspected in order to find any sources of lead.

Q: **The doctor says my child's lead level is between 10 and 14 μg/dL. What does this mean? Has my child been damaged?**

A: Studies of the effects of lead on large groups of young children show that lead can be harmful at these blood levels. It is important to remember, however, that these effects are seen in studies of large groups of children and do not mean that every child will have the same problems. To help your child grow up healthy and smart, make sure she or he receives the proper foods, adequate shelter, and plenty of love.

This document was prepared by the U.S. Department of Health & Human Services, Public Health Service, Centers for Disease Control, Atlanta, Georgia, for distribution by the National Lead Information Center. 10/92

The National Lead Information Center is operated by the National Safety Council with funding from the Environmental Protection Agency, the Centers for Disease Control, the Department of Housing and Urban Development, and the Department of Defense.

FIGURE 2-2. National Lead Information Center fact sheet concerning parental questions about lead poisoning.

can no longer support the foot due to nerve damage. This has been called "lead palsy" or "painters palsy" and frequently occurs from a chronic exposure to lead.

Blood

When lead is absorbed into the bloodstream, it will attach to the red blood cell (erythrocytes). Lead interferes with the production of both hemoglobin and blood cells, resulting in anemia. The affects of anemia are well-known and include fatigue, exhaustion, and pale coloring. Lead is also known to reduce hepatic heme, upon which enzyme formation occurs, resulting in a lowering of the body's ability to detoxify pollutants. This can cause numerous health problems that may, upon initial examination, seem unrelated to lead exposure.

Adverse impacts of the heme formation pathway and on vitamin D and calcium metabolism, all of which have far-reaching physiological effects, have been documented at blood lead levels of 15–20 μg/dl in children. At levels around 40 μg/dl, the effects on heme synthesis increase in number and severity (e.g., reduced hemoglobin formation).

Gastrointestinal Tract

The gastrointestinal tract includes the stomach, small and large intestines, and the rectum. Since many lead exposures are a result of ingestion, the gastrointestinal tract is one of the first target organs lead encounters. Symptoms of overexposure include constipation and acute abdominal pain, commonly called "lead colic." These symptoms usually occur at blood lead levels of greater than 80 μg/dl.

Kidneys

The major function of the kidneys is to filter substances in the bloodstream that may prove harmful to the body. As blood passes through the kidneys, impurities, such as lead, are absorbed by the organ and excreted from the body as urine. Chronic exposure to lead eventually interferes with this filtering process by altering the metabolism of the kidney. Kidney damage is irreversible and has been recognized as a disease affecting many lead workers.

More than 150 years ago a researcher described a young man who died of renal disease without some of the characteristic presentations of the disease. This individual happened to be a plumber with a significant amount of occupational lead exposure. This was one of the first suggestions of a relationship between lead and diseases of the kidney. By the end of the 1800s, additional research firmly established the existence of lead *nephropathy,* or disease of the kidneys caused by lead exposure. In the decades that followed, there seemed to occur a time during which this relationship was ignored. This may have been because occupational exposure often caused characteristic symptoms in acute conditions. Among these conditions were colic, palsy, and encephalopathy. It was this thinking with regard to binary outcomes that either a person is lead poisoned and develops characteristic symptoms, or he or she is not; that may have contributed to the oversight of previously established medical facts.

There were several somewhat concurrent observations that reinforced the previous documentation of the relationship between lead exposure and compromised renal function. First was the Queensland study. This observation took place in Queensland, Australia, and although it involved children, it firmly established the relationship between lead and renal disfunction in children who had been poisoned with lead carbonate.

These poisoned children were monitored closely in subsequent years and there was an increase in the number of deaths due to renal failure. In 1954, Henderson found that 54 percent of the survivors had renal disease and through the examination of medical records found evidence that 67 percent of those who had died before they were 40 had some record of renal disease.[7]

Evidence establishing the effects of lead on renal function in adults also came from observations in the Southwest of a high occurrence of renal failure in individuals who partook of illegally distilled whiskey. Renal disfunction of this kind has been termed the "moonshine malady." Illegally distilled whiskey in this area contains high levels of lead, which is leeched from the apparatus utilized in the distillation process. The moonshine malady is frequently accompanied by gout, which is not a common finding in renal diseases of other types, but is frequently found in cases of chronic lead intoxication. It was determined that most of the gouty patients in a Veterans Administration Hospital in Alabama and were either manufacturers or drinkers of homemade whiskey.

One last general area of observation involved occupational lead exposure. As stated earlier, the relationship of lead poisoning and occupational exposure was established prior to the turn of the century but subsequently forgotten or ignored. In the early part of the twentieth century studies were done on workers in battery manufacturing plants, smelting operations, enameling occupations, and paint manufacturers. Like asbestos, negative health effects of occupational lead exposure at first met with a great deal of skepticism. However, over the past years a lot of progress has been made as hundreds of occupational exposure situations have been identified and the compromising health effects have been fully established. Seventeen percent of Rumanian patients were found to have chronic nephropathy following 10 years of lead exposure. Twenty-three percent of exposed employees in a lead smelting operation had contracted kidney disease. Excess mortality due to kidney disfunction has been documented in exposed workers in lead battery plants, primary smelters, refineries, recycling plants, and secondary smelters.

OTHER INDICATIONS

Gout

The association between lead and gout was recognized back during the beginning of the eighteenth century. It was first recognized following a epidemic of lead poisoning that resulted from people drinking lead contaminated cider. The high number of cases of gout in patients with renal failure and elevated blood lead levels is unique since it is not usually associated with renal failure.

Lead and Blood Pressure

Greatly elevated blood pressure has been documented for many years as a manifestation of acute lead intoxication and extreme constriction of the vessels occur with blood lead levels of 70–200 μg/dl.

The controversy centers now, however, on the effects on blood pressure of lower blood lead levels. In the areas of occupational and environmental exposure, "relativity" is the byword.

The amount of lead in the ambient environment probably, in itself, results in modern men and women having blood lead levels far in excess of preindustrial age and certainly prehistoric humans. Studies utilizing isotopic techniques have shown that present day blood lead levels that we consider normal are probably 200–500 times greater than preindustrial man. So, in this context, we are existing while harboring extraordinary historically high stores of lead in our bodies. If lower levels of lead intoxication can affect blood pressure in the range of a few points, then in a global sense, the question is, What effect will this increase have upon the statistical probability of marked increases in heart problems in modern times on large populations?

Research has attempted to predict the health compromising effects of the increase in blood pressure caused by elevated levels of blood lead. Through statistical manipulation it may be possible to predict the change in blood pressure that would result from a specific change in blood lead levels, and subsequent to that, to use other medical data and studies to calculate the changes in cardiovascular events that would result from these changes.

There was a 37 percent decrease in the mean levels of blood lead in white males from 1976 to 1980. This decrease is believed to have occurred as a result of a decrease in ambient (or environmental) exposure, since ambient lead levels have fallen dramatically during this period due to the decrease in the use of leaded gasoline. On the basis of this 37 percent decrease, it was predicted that there would be significant decreases in myocardial infarction, stroke, and death from all heart related events.

Event	*Predicted Decrease*
Myocardial Infarction	4.7%
Stroke	6.7%
Death from all causes	5.5%

Since the effect in this case is restricted to events related to blood pressure, it can be implied that the overall effect would be significantly greater if one takes into consideration some of the other health-compromising effects of lead. However, given the great many other confounders that would also affect blood pressure and related cardiac events, it would be impossible to accurately assess the validity of this prediction.

Effects on the Visual System

Since lead is a neurotoxin, it is reasonable to expect that exposure to lead in amounts sufficient to increase the body content of lead would affect the visual neurological apparatus, since the visual system is an exceptionally sensitive target for neurotoxins. And, a frequent observation in acute exposure is that the optic nerve becomes inflamed and severe compromise of vision is noted.

There have not been a great deal of studies accomplished on the effects of chronic exposure on the visual system in humans. In studies accomplished on monkeys and rats, however, significant visual problems were encountered following chronic lead exposures. In human studies visual impairment was found to be induced by low to medium levels of lead from chronic exposure. In workers who have been exposed to lead stearate on their jobs, 28 percent had a selective loss of central vision and a significant reduction in their visual field. It can be implied that the effects of lead in the various target organs of the toxin in adults take place in a continuous mode and some effect may occur even at low levels of intoxication.

Reproductive System

Lead affects the reproductive organs of both men and women. Altered testicular function among men and effects on the female organs has been observed at levels over 40 µg/dl. In men, chronic lead poisoning may create abnormal sperm cells and/or a decrease in sperm mobility. It may also lead to a reduced sex drive. In woman, lead poisoning may manifest itself in abnormal menstrual cycles, decreased fertility, and, according to some researchers, a marked increase in miscarriages and stillbirths.

Potential Carcinogenicity

One of the biggest questions related to the health effects of lead is whether exposures can cause cancer. Both the International Agency for Research on Cancer (IARC) and the U.S. Environmental Protection Agency (EPA) have concluded that lead is a carcinogen (cancer-causing substance) based on laboratory tests performed on animals. However, both organizations concluded that human data are inadequate to refute or demonstrate the potential carcinogenicity of lead exposure for humans. The IARC classified lead and inorganic lead compounds as an IARC Group 2B substance, a "possible human carcinogen." The EPA assigned lead and inorganic lead compounds as B2 substances—probable human carcinogens. The EPA's classification was based on numerous scientific studies and was endorsed by EPA's Science Advisory Board.

FETAL EXPOSURE

Lead that is circulating in a pregnant woman's blood can readily cross the placenta during pregnancy and thereby reach the developing fetus. After birth, maternal blood lead can concentrate in the mother's breast milk and continue to reach the infant during nursing.

Exposure of a fetus to lead levels as low as 10–15 μg/dl are associated with developmental effects during the first two years of life. These include effects on memory, learning, problem solving, and the beginnings of verbal communication. Research studies suggest that increased levels of in utero exposure to lead may result in other effects including preterm deliveries, low birth weight, miscarriages, and stillbirths.

It is important to note that the harmful effects of lead on a fetus occur at maternal blood lead levels significantly lower than levels permitted under OSHA's workplace standard for lead exposure.

ADULT VERSUS CHILD EXPOSURES

The daily exposure of a given amount of lead to both an adult and a child would pose a greater danger to the child. Consider that:

- A child's nervous system is undergoing rapid development and is particularly susceptible to permanent damage.
- A child breathes more air and consumes more food per unit of body measure (weight and body surface area) compared to an adult. Therefore the amounts of inhaled and dietary lead are proportionally greater for a child than an adult.
- Children absorb approximately 50 percent of the lead they ingest while adults typically absorb about 10 percent.
- The turnover rate of minerals in a child's bone is greater than an adult, resulting in a greater availability of lead to the child's developing body systems. Children also retain a greater fraction of the lead they absorb into their rapidly growing skeletal systems and soft tissues.

Furthermore, these disproportionate risks to children occur at a time when children are most in need of optimal nutrition but when nutrition deficiencies are most likely to occur. Deficiencies in iron, calcium, and other dietary elements enhances lead uptake and retention in children and increases their risk of lead poisoning.

IMPLICATIONS FOR THE LEAD-BASED PAINT ABATEMENT INDUSTRY

Unlike asbestos or radon levels, which are difficult to detect in exposed individuals and which may cause health problems many years after exposure, exposures to lead are quickly measurable and the health consequences can be immediate and severe. The risks of lead exposure are not based on theoretical calculations; they are well known from studies of children themselves and are not extrapolated from data on laboratory animals or high-dose occupational exposures.[8]

Contractors are well served to keep in mind that the units of measuring blood lead—millionths of a gram in a deciliter—are quite low and that careful attention to detail in performing lead-based paint abatements and cleanups will be required to protect occupants, especially children, who will occupy abated areas and to prevent these occupants as well as their abatement workers from being exposed to conditions that may result in measurable increases in blood lead levels.

Suggested Readings

Audesirk, G., and T. Audesirk. Effects of chronic low level lead exposure on the physiology of individually identifiable neurons. *NeuroToxicology* 15–25 (1983).

Batuman, V., E. Landy, J. Maesaka, and R. Wedeen. Contribution of lead to hypertension with renal impairment. *New England J. Med.* 309(1):17–21 (1983).

Batuman, V., J. Maesaka, B. Haddad, E. Tepper, E. Landy, and R. Wedeen. The role of lead in gout nephropathy. *New England J. Med.* 304(9):520–523 (1981).

Beck, L. Requiem for gouty nephropathy. *Kidney International* 30:280–287 (1986).

Bellinger, D., H. Needleman, A. Leviton, C. Waternaux, B. Rabinowitz, and M. Nichols. Early sensory-motor development and prenatal exposure to lead. *Neurobehavorial Toxicology and Teratology* 6:387–402 (1984).

Bennett, W. Lead nephropathy. *Kidney International* 28:212–220 (1985).

Boscolo, P., and M. Carmignani. Neurohumoral blood pressure regulation in lead exposure. *Environ. Health Perspectives* 78:101–106 (1988).

Cavalleri, A., F. Trimarchi, C. Gelmi, A. Baruffini, C. Minoia, G. Biscaldi, and G. Gallo. Effects of lead on the visual system of occupationally exposed subjects. *Scand. J. Work Environ. Health* 8:148–151 (1982).

Colleoni, N., and G. D'Amico. Chronic lead accumulation as a possible cause of renal failure in gouty patients. *Nephron* 44:32–35 (1986).

Cooper G., J. Suszkiw, and R. Manalis. Heavy metals: Effects on synaptic transmission. *NeuroToxicology* 5(3):247–266 (1984).

Goldstein, G., A. Asbury, and I. Diamond. Pathogenesis of lead encephalopathy—uptake of lead and reaction of brain capillaries. *Arch. Neurol.* 31:382–389 (1974).

Grandjean, Philippe, B. Jensen, S, Sando, P. Jorgen, S. Antonsen. Delayed blood regeration in lead exposure: An effect on reserve capacity. *Am. J. of Pub. Health* 79:1385–1388 (1989).

Harlan, W. R., The relationship between blood lead levels to blood pressure in the U.S. population. *Environ. Health Perspectives* 78:9–13 (1988).

Harlan, W., J. Landis, R. Schmouder, N. Goldstein, and L. Harlan. Blood lead and blood pressure. *JAMA* 253(4):530–534 (1985).

Kirkby, H., and F. Fyntelberg. Blood pressure and other cardiovascular risk factors of long-term exposure to lead. *Scan. J. Work Environ. Health* 11:15–19 (1985).

Kromhout, C. Blood lead and coronary heart disease risk among elderly men in Zutphen, the Netherlands. *Environ. Health Perspectives* 78:43–46 (1988).

Mushak, P., J. Davis, A. Crocetti, and L. Grant. Review: Prenatal and postnatal effects of low-level lead exposure: Integrated summary of a report to the U.S. Congress on childhood lead poisoning. *Environ. Research* 50:11–36 (1989).

Needleman, Herbert L., M. Rabinowitz, A. Leviton, S. Linn, and S. Schoenbaum. The relationship between prenatal exposure to lead and congenital abnormalities. *JAMA* 251:2955–2959 (1984).

Needleman, H., A. Schell, D. Bellinger, A. Leviton, and E. Allred. The long-term effects of exposure to low doses of lead in childhood. *New England J. Med.* 322(2):83–88 (1990).

Orssaud, G., J. Claude, T. Moreau, J. Lellough, B. Juguet, and B. Festy. Blood lead concentration and blood pressure. *Brit. Med. J.* 290:54 (1985).

Parkinson, D., M. Hodgson, E. Bromet, M. Dew, and M. Connell. Occupational lead exposure and blood pressure. *Brit. J. Ind. Med.* 44:744–748 (1987).

Pinto de Almeida, A., F. Carvalho, A. Spinola, H. Rocha. Renal dysfunction in Brazilian lead workers. *Am. J. Nephrol.* 7:455–458 (1987).

Pirkle, J., J. Schwartz, J. Landis, and W. Harlan. The relationship between blood lead levels and blood pressure and its cardiovascular risk implications. *Am. J. Epidemiology* 121(1):246–258 (1985).

Pocock. S., A. Shaper, D. Ashby, H. Delves, and B. Clayton. The relationship between blood lead, blood pressure, stroke, and heart attacks in middle-aged British men. *Environ. Health Perspectives* 78:23–30 (1988).

Ritz, E., J. Mann, and M. Stoeppler. Lead and the kidney. *Adv. Nephrol.* 17:241–274 (1988).

Savitz, D., E. Whelan, and R. Kleckner. Effect of parent's occupational exposures on risk of stillbirth, preterm delivery, and small-for-gestational age infants. *Am. J. Epidemiology* 129:1201–1218 (1989).

Schwartz, J. The relationship between blood lead and blood pressure in the NHANES II Survey. *Environ. Health Perspectives* 78:15–22. (1988).

Schwartz, J. Lead, blood pressure, and cardiovascular disease. In *Human Lead Exposure,* H. Needleman, ed. CRC Press, Boca Raton, FL (1992).

Schwartz, J. Low level health effects of lead: Growth, developmental, and neurological disturbances. In *Human Lead Exposure,* H. Needleman, ed. CRC Press, Boca Raton, FL (1992).

Schwartz, J., and D. Otto. Blood lead, hearing thresholds, and neurobehavorial development in children and youth. *Arch. Environ. Health* 42(2):153–159 (1987).

Schwartz, J., and D. Otto. Lead and minor hearing impairment. *Arch. Environ. Health* 46(5):300–304 (1991).

Silbergeld, E., and H. Adler. Subcellular mechanisms of lead neurotoxicity. *Brain Research* 148:451–467 (1978).

Tennekoon, G., C. Aitchison, J. Frangia, D. Price, and A. Goldberg. Chronic lead intoxication: Effects on developing optic nerve. *Ann. Neurol.* 5:558–564 (1979).

Wedeen, R. Blood lead levels, dietary calcium, and hypertension. *Ann. Int. Med.* 102(3):403–404 (1985).

Wedeen, R. P. Bone lead, hypertension, and lead nephropathy. *Environ. Health Perspectives* 78:57–60 (1988).

Wedeen, R. P. Lead, the kidney, and hypertension. In *Human Lead Exposure,* H. Needleman, ed. CRC Press, Boca Raton, FL (1992).

Wibberly, D. G., A. Khera, J. Edwards, and D. Rushton. Lead levels in human placentae from normal and malformed births. *J. Med. Genetics* 14:339–345 (1977).

Notes

1. U.S. Environmental Protection Agency. *Air Quality Criteria for Lead,* EPA report no. EPA-600/8-83/018F Research Triangle Park, NC: Office of Health and Environmental Assessment, Environmental Criteria and Assessment Office, June 1986, addendum, September 1986. (Cited in ATSDR 1990.)
2. Agency for Toxic Substances and Disease Registry. *Toxicological Profile for Lead,* ATSDR/TP-88/17. Prepared by Syracuse Research Corporation under contract No. 68-C8-0004, June 1990.
3. P. A. Succop, E. J. O'Flaherty, R. L. Bornschein, et al. A kinetic model for estimating changes in the concentration of lead in the body of young children. In *International Conference: Heavy Metals in the Environment,* Vol. 2, S. E. Lindberg and T. C. Hutchinson, eds. Edinburgh, UK: CEP Consultants, Ltd., 1987, p. 289–291. (Cited in ATSDR 1990.)
4. Centers for Disease Control. *Preventing Lead Poisoning in Young Children.* October 1991, p. 2.
5. National Academy of Sciences. *Measuring Lead in Infants, Children, and Other Sensitive Populations.* National Academy Press, 1993.
6. Centers for Disease Control, p. 9.
7. D. A. Henderson. A follow-up of cases of plumbism in children. *Austral. Ann. Med.* 3:219 (1954).
8. Centers for Disease Control, p. 7.

3

Populations At Risk

Vincent M. Coluccio, Dr.P.H.

RISK LEVELS

The Centers for Disease Control (CDC) uses the term *elevated blood lead levels* (EBL) to classify a person whose blood lead level indicates the need for interventions such as medical treatments and/or inspections and abatement of environmental sources of lead. Blood lead levels are used as guidelines for determining the appropriate intervention for different levels of lead absorption. Table 3-1 shows CDC's periodic lowering of the EBL definition for children over the past few decades as medical research revealed poisonous effects of lead at lower and lower blood levels.

Young children and fetuses are at most risk for lead poisoning. In October of 1991, the CDC published a document entitled *Preventing Lead Poisoning in Young Children,* which lowered its definition of elevated blood lead level from 25 μg/dl to 10 μg/dl for preschool children and established a multitiered intervention strategy based on blood lead levels.[1] Table 3-2 outlines this approach.

To put these blood lead values in perspective, consider that the 10 μg/dl action level is low by comparison with levels that are associated with observable toxic reactions and that were used as action levels in the 1960s and 1970s (50–80 μg/dl) to intervene in protecting children. Furthermore, studies of bone samples of preindustrial North American Indians and other populations indicate that body burdens of lead in the general population today are 300–500 times greater than preindustrial background concentrations.[2] The average blood lead level in the United States in the early 1990s was approximately 5 μg/dl, with a few percentage points higher average for urban populations and a few points lower average for rural populations.

TABLE 3-1. Blood Lead Levels Considered Elevated by the Centers for Disease Control and Public Health Service

Year	Blood Lead Level (μg/dl)
1970	40
1975	30
1985	25
1991	10

TABLE 3-2. CDC Interpretation of Blood Lead Level Results for Children

Class	Blood Lead Concentration (μg/dl)	Comment
I	<9	Not considered lead poisoned.
IIA	10–14	Many children (or a large proportion of children) with blood lead levels in this range should trigger community-wide lead poisoning prevention activities. Children in this range may need to be rescreened more often.
IIB	15–19	Child should receive nutritional and educational interventions and more frequent screening. If the blood lead level persists in this range, environmental investigation and intervention should be done.
III	20–44	Child should receive medical evaluation; environmental evaluation and remediation may be warranted. Child may need pharmacologic treatment of lead poisoning.
IV	45–69	Child needs both medical and environmental interventions, including chelation therapy.
V	>70	Medical emergency. Medical and environmental management must begin immediately.

Source: CDC, *Preventing Lead Poisoning in Young Children*, 1991.

CHILDHOOD LEAD POISONING

Estimated Numbers

Comprehensive data on the extent of lead poisoning in children and fetuses is not available at this time, although studies are currently underway to develop these data in many regions of the country. One reason for this lack of data is that the majority of children with elevated blood lead levels appear well and are not identified through routine medical examinations. Furthermore, physicians and laboratories have been required to report only blood lead levels above 25 μg/dl, more than twice the level now endorsed by the CDC.

The best estimates on the extent of childhood lead poisoning come from the National Health and Nutrition Examination Survey II (NHANES II), conducted by the CDC's National Center for Health Statistics. The findings and conclusions of NHANES II are based on data collected from 1976 through 1980, extrapolated to 1984 population, and predicated on assumptions, some of which tend to overestimate and some of which underestimate the number of lead poisoning cases. Furthermore, the NHANES data are limited to black and white children (not Hispanic and "other race" origins) between the ages 5 months to 6 years and residing in metropolitan areas (referred to as Standard Metropolitan Statistical Areas, or SMSAs).

In balancing all sources of overestimates and underestimates inherent in the NHANES II methods, the federal Agency for Toxic Substance and Disease Registry (ATSDR) believes that the NHANES II projections of childhood lead poisoning cases are probably close to the actual values.[3] Some key findings of the NHANES II as summarized by ATSDR are provided below:

> About 2.4 million white and black metropolitan children, or about 17 percent of such children in U.S. SMSAs, have blood lead levels above 15 μg/dl and therefore are exposed to environmental sources of lead at concentrations that place them at risk of adverse health effects. This number approaches 3 million black and white children if extended to the entire U.S. child population. If the remaining racial categories are included in these totals, a total of between 3 and 4 million U.S. children may be affected.

It is noteworthy that the NHANES II estimates cited above refer to children with blood lead levels above 15 μg/dl. Since the NHANES II data were reported in 1988, the CDC has redefined childhood lead poisoning as a blood lead level of 10 μg/dl.

The number of fetuses exposed to lead in utero has also been estimated:

> In a given year, an estimated 400,000 fetuses (within SMSAs alone) are exposed to maternal blood lead levels of more than 10 μg/dl and therefore at risk for adverse health effects. This number pertains to a single year; the cumulative number of children who have been exposed to undesirable levels of lead during their fetal development is much greater. . . .

While NHANES II supported the conventional wisdom that the highest prevalence of elevated blood lead levels is found among children residing in inner-city, impoverished areas, it also documented that no economic or social subgrouping of children is exempt from lead exposure and lead poisoning.

Screening

The traditional goal of childhood lead poisoning screening programs has been to identify asymptomatic lead poisoned children (those without obvious signs and symptoms of disease) and then to intervene as quickly as possible to abate sources of lead exposure and reduce blood lead levels. An additional benefit of screening programs is that abatement results in the prevention of lead poisoning for other children who would have been exposed to those sources in the future.[4]

Screening must be done using a blood lead test. In past years, the erythrocyte protoporphyrin (EP) test was used to screen children for lead poisoning but, as discussed in Chapter 2, the EP test is not sensitive enough to identify more than a fraction of small children whose blood lead levels are between 10 μg/dl and 25 μg/dl. Therefore the CDC recommends that all children be screened using blood lead tests as the primary screening method.[5] The exception to this "universal screening requirement" is for communities that can verify through testing of a large percentage of children that a lead poisoning problem does not exist.

Blood lead tests involve obtaining blood either through a venous puncture (i.e., blood drawn from a vein) or on capillary blood samples obtained from a finger stick, which draws a drop of blood. Although capillary samples are easier, quicker, and less expensive to obtain, they can easily be contaminated by small lead dust particles on a child's skin and result in a "false positive" test. However, obtaining capillary specimen is often more feasible than obtaining a venous specimen, particularly on a small child.

Sources of Lead Exposure

The ATSDR estimated the number of children exposed to lead from each of six major sources: lead-based paint, gasoline, stationary sources, dust/soil, water; and food. Due to the complex and interrelated exposure pathways from these sources it is not possible to establish exact counts of children exposed to each source. Again, ATSDR's estimates are based on available information and a defined set of assumptions and contain considerable discussion on the limitations in the precision of the calculations. Its estimates of potentially exposed children and estimates of actually exposed children (the former necessarily greater than the latter in each case) are summarized below by source category:

Lead-based paint: Children under 7 years old who are potentially exposed to paint surfaces containing lead concentrations of 0.7 mg/cm^2 or higher number about 13.6 million, of whom about 5.9 million live in the oldest, highest lead paint residential units. Of these children, about 4.4 million live in U.S. SMSAs. Children in deteriorated, old housing number 1.8–2.0 million.

Children living in old and deteriorated housing and estimated to have blood lead levels above certain selected criterion values due mainly to paint lead exposure amount to: 1.2 million at greater than 15 μg/dl, 0.5 million at greater than 20 μg/dl, and 0.2 million at greater than 25 μg/dl.

Gasoline: Children potentially exposed to lead from combusted gasoline and residing in the 100 largest cities of the United States are estimated to total about 5.6 million.

Children 13 years old and younger who are sufficiently exposed to gasoline lead to show declines in blood lead levels in response to the phasedown of leaded gasoline amount to: 1.6 million falling below 15 μg/dl and about 0.6 million falling below 20 μg/dl.

Stationary sources: For stationary sites, namely primary and secondary smelters, the potential exposure estimate is about 230,000 children. Of this number, up to 13,000 children will have blood lead levels above approximately 20 μg/dl.

Dust and soil: For lead in dust and soil, it is necessary to use source-specific numbers for the generators of this source, that is, paint lead in older housing, lead fallout from leaded gasoline combustion, and stationary emissions. With regard to the potential exposure of children, these primary source surrogates for dust and/or soil lead amount to a range of 5.9–11.7 million children. Actual exposures of children to soil/dust lead sufficient to raise blood lead levels to the range of toxicity risk cannot be estimated at this time, but the actual number may be considerable.

Drinking water: Because of lead in old residential plumbing, 1.8 million children under 5 years old and 3.0 million children 5–13 years old are potentially exposed to lead; for new residences (less than 2 years old), the corresponding estimates of children are 0.7 and 1.1 million, respectively.

Some actual exposure to lead occurs for an estimated 3.8 million children whose drinking water lead level has been estimated at greater than 20 parts per billion.

The EPA in a recent study estimated that 241,000 children under 6 years old have blood lead levels above 15 μg/dl because of elevated concentrations of lead in drinking water. Of this number, 100 have blood lead levels above 50 μg/dl, 11,000 have blood lead levels between 30 and 50 μg/dl, and 230,000 have blood lead levels between 15 and 30 μg/dl.

Food: Most children under 6 years of age in the U.S. child population are potentially exposed to lead in food at some level. Actual exposure to enough lead in food to raise blood lead levels to an early toxicity risk level has been estimated to impact as many as 1 million children. This is an overestimate arising from the fact that it is based on food lead data for the 1970s, when levels were higher than for more recent years. Lead in food has been reduced significantly in recent years and contributes less to body burdens in the United States than in the past.

Of all potential sources of lead exposure, lead-based paint is associated with the greatest number and severest forms of childhood lead poisoning.

Treatment

The best treatment for lead poisoning is prevention. The management of a child's lead poisoning potential involves eliminating the lead source from the child's environment, providing general pediatric care with focus on lead detection, educating families on the hazards of lead paint, and recognizing potential sources of lead contamination. A parent must also understand that nutritional deficiencies—especially calcium, iron, and phosphorous—along with an increase in dietary fat elevates the absorption of lead.

According to the CDC protocols discussed above, children with blood lead levels between 10 μg/dl and 19 μg/dl need followup and repeat blood lead screening but do not need medical evaluation. All children with confirmed venous blood lead levels equal to or greater than 20 μg/dl require medical evaluation.

When serious lead poisoning cases occur a drug therapy called *chelation* (pronounced "key-lation") is warranted. Chelation is a process in which a drug binds itself to lead and makes that lead easy to excrete. The drugs used in chelation therapy include British Anti-Lewisite (BAL), Cuprimine (D-penicillamine), and Calcium Disodium Versenate ($CaNa_2$-EDTA). In addition a new oral chelating agent known as Succimer was approved for use by the U.S. Food and Drug Administration in 1991.

The CDC advises chelation therapy for children with blood lead levels greater than or equal to 45 μg/dl, and different clinics and programs use different protocols to medically manage children with blood lead levels from 25–44 μg/dl. Chelation therapy is required in a relatively small number of cases and usually requires hospitalization for up to five or more days for chelation and posttreatment testing. Certain risks and limitations that are inherent to chelation therapy are described by CDC.

Chelation has little effect if the patient returns to a lead-contaminated environment. If the parent does decide on abatement, extra precautions must be taken to protect the child from reexposure. A child living at home during an abatement can have blood lead measurements of 30–50 μg/dl within a few days or weeks. The authors are aware of instances in which children return many times for chelation therapy because the lead in the environment in which the children lived was not properly abated. Chelation therapy should never be used as a substitute for preventing lead exposures and/or abating lead hazards.

CASE STUDY: NEW YORK CITY

New York City has an enormous problem involving the interwoven issues of childhood lead poisoning, an aging housing stock containing lead-based paint, and a shortage of affordable housing. A review of the city's dilemma is instructive in that it illustrates the need for development of lead hazard risk assessment and interim control measures discussed in this book and the cultivation of a workforce to competently provide these services.

New York City has the unfortunate distinction of having the greatest number of reported childhood lead poisoning cases compared to other municipalities nationwide, with approximately 650 new cases being reported each year and 4,000 cases under medical and/or environmental management.[6] These figures were reported for 1990, are based on a lead poisoning case definition of greater than 25 μg/dl, and account for about 8 percent of the nationwide caseload of lead poisoned children. The 650 cases were identified largely through the blood lead screening of 350,000 of New York City's 700,000 children under the age of seven.

The following is a profile of New York City's childhood lead poisoning caseload:

- 50 percent of the cases are black children, 30 percent Latino; 20 percent white and "other" races.
- 50 percent of the cases live in a nearly contiguous patch of neighborhoods referred to as the "lead belt," extending from the borough of Brooklyn into Queens, upper Manhattan, and lower Bronx. Lead belt neighborhoods are economically depressed and have an aging, deteriorating, and mostly lead-painted housing stock.
- More than 80 percent of the lead poisoned children live in homes with deteriorating lead-based paint.

Each report of a lead poisoned child in New York City triggers a lead-based paint inspection of the child's residence and medical case management by the city's Health Department. The city orders abatement of deteriorating and/or mouthable lead-based paint (paint on a protruding surface that is within reach of a child) that it identifies during these inspections. In 1991 the city inspected 3,300 dwelling units; 513 abatements (consisting mostly of sheetrocking and repainting) were completed at the average cost of $2,500 per abatement.[7]

The city recently adopted the CDC lead poisoning definition of 10 μg/dl and lowered the blood lead trigger for medical and environmental case management from 25 μg/dl to 20 μg/dl. Each year, an estimated 6,000 to 10,000 children in New York City have blood lead levels exceeding 20 μg/dl and another 65,000 children exceed 10 μg/dl.[8] Figure 3-1 graphically displays the number of reported childhood lead poisoning cases in New York City from 1970 through 1991, along with the effect of an increase in caseload each time the case definition (in μg/dl) was lowered. Note that a plot of the anticipated increase of 6,000 to 10,000 new cases requiring medical evaluations and environmental interventions will extend the height of the vertical axis of the chart in Figure 3-1 at least twofold.

New York City was among the first municipalities to regulate lead-based paint. In 1960 the city banned the use of lead-based paint in residences,[9] and in 1982 the City passed Local Law 1 requiring owners of multiple dwellings to remove or

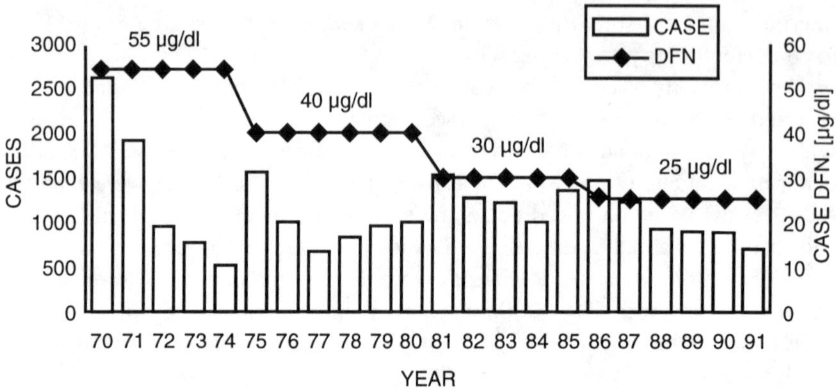

FIGURE 3-1. Lead poisoning cases and case definition in New York City, 1970–1991.

cover lead-based paint in all units where children under the age of 7 reside, regardless of the condition or location of such paint. Local Law 1 also establishes a presumption that any deteriorating paint present in a pre-1960 multiple dwelling where children live is lead-based paint, and that such condition is a violation and must be corrected by the building owner. Where the owner does not correct the violation, the city may order and pay for an emergency repair of the violation and seek payment for the repairs from the owner.

In 1992 the city's Department of Housing Preservation and Development oversaw 1,767 Local Law 1 projects (mostly scrape-and-sweep removals of small patches of deteriorating lead-based paint) at the average cost of $618.[10]

In 1985 a lawsuit was brought against the city by the New York City Coalition to End Childhood Lead Poisoning (NYCCELP).[11] Among the charges filed by NYCCELP was that the city was not enforcing Local Law 1; that the city was ordering lead abatements only of those dwellings occupied by lead-poisoned children instead of all multiple dwelling units where children reside as required by Local Law 1; that the abatements were only of deteriorating lead-painted surfaces instead of all lead-painted surfaces regardless of condition in those units; and that the city lacked proper abatement and safety guidelines.

In 1992 the New York State Supreme Court ruled in favor of NYCCELP, finding that the city must enforce Local Law 1 abatement provisions, as well as adopt safety regulations and implement other changes to upgrade its response to lead poisoning.

The magnitude of the financial costs for implementing Local Law 1 and the associated adverse impacts on the city's affordable housing impede the city's ability to comply with the law, a position that has put the city in contempt of court. Within the city are approximately 2.2 million pre-1960 apartments (most of which are presumed to contain lead-based paint). If the city were to require Local Law 1 abatements of only 1 million of these units in blocks of 10,000 units per year, and the average cost of abatement is $15,000 per unit, total costs would exceed $150 million per year for the next 100 years. Unfortunately this scenario understates the problem in three major ways: (1) an experienced and competent workforce is not in place to safely abate lead-based paint in 10,000 units per year; (2) abatement of nonhazardous lead paint (for example, intact ceiling paint covered with nonleaded paint), which would be required under Local Law 1 in many apartments and implemented by an unprepared workforce, will very likely cause widespread hazardous lead conditions where none existed; and (3) the actual costs for the full and proper abatement of many dwelling units may exceed $15,000 per unit.[12]

New York City officials have proposed modifying Local Law 1 to replace the mandatory abatement requirements with a multitiered abatement approach. This approach was recommended by a panel of consultants hired by the city[13] and it calls for performing lead hazard risk assessments of dwelling units and rating the units according to the degree of risk present from Level 1 (least risk) to Level 4 (greatest risk). The risk assessment takes into account the extent of lead paint deterioration and dust lead levels present in a unit, followed by hazard controls (from in-place management of lead-based paint to full abatement) that are appropriate for the degree of risk present rather than full abatement as required under Local Law 1. The city's proposed risk assessment strategy is outlined in Table 3-3.

The newly established protocols for risk assessment and interim control strategies are discussed in this text, along with a discussion of their benefits and limitations. It is important to note that many questions regarding the technical soundness, cost-effectiveness, and future commitment of resources needed to support these strategies have yet to be answered. It is therefore vital to the attainment of our lead poisoning prevention goals that upcoming risk assessment/interim control initiatives by all parties, the city of New York included, be closely studied and evaluated at their earliest phases and that necessary corrections be made to enhance their chances of success.

Notes

1. Centers for Disease Control. *Preventing Lead Poisoning in Young Children.* Atlanta: U.S. Department of Health and Human Services/Public Health Service, October 1991.
2. Agency for Toxic Substance and Disease Registry. *The Nature and Extent of Lead Poisoning in the United States: A Report to Congress,* July 1988.
3. Ibid.
4. Centers for Disease Control, p. 39.
5. Ibid.
6. NYC DOH. Dr. Andrew Goodman, NYC Mayor's Lead Paint Poisoning Prevention Advisory Committee presentation of February 18, 1992.
7. NYC Mayor's Advisory Committee to Prevent Childhood Lead Poisoning. Draft Final Report, June 1993.

TABLE 3-3. New York City's Proposed Lead-Based Paint Abatement Strategy

	Lead Levels in Dust (μg/ft^2)				
	Floor	Sill	Well[a]	Abatement Measures	City-Estimated Costs[b]
Level 1	<200	<500	<800	Wet scrape and repaint.	$ 1,200
Level 2	200–1000	500–2500	800–8000	Delead friction surfaces. Encapsulate chewable surfaces. Wet scrape and repaint peeling areas.	$ 2,980
Level 3	>1000	>2500	>8000	Replace windows. Encapsulate chewable surfaces. Enclose other deteriorated surfaces. Paint floors. Wet scrape and repaint peeling areas.	$ 7,913
Level 4	In dwellings where a lead-poisoned child lives, the health commissioner may order additional abatement measures as necessary.				Up to $15,000
All levels: work area preparation, clean-up with HEPA vacuum and wash, postabatement dust tests, annual monitoring for 2 years.					

Source: NYC Mayor's Advisory Committee, Draft Report (Fourth Revision), August 1993.

[a]Lead-based paint hazard ratings would not be based solely upon the window well reading.

[b]These figures do not include the costs of testing or disposal.

8. NYC Mayor's Advisory Committee, Draft Report (Fourth Revision), August 1993.
9. New York City Health Code, Section 173.13.
10. NYC Mayor's Advisory Committee.
11. *New York Coalition to End Lead Poisoning v. Koch,* IAS Part 25, 1985.
12. A client of ATC Environmental Inc. received contractor bids ranging from $29,000 to $148,000 to fully abate lead-based paint conditions in one upscale apartment unit in Manhattan.
13. D. E. Jacobs. "Lead Abatement in New York City: A Discussion on Prioritizing Risk." Prepared for the NYC DOH and HPD, February 15, 1992.

4

Key Federal Regulations

Vincent M. Coluccio, Dr.PH.

David Chambers

LEAD BASED PAINT POISONING PREVENTION ACT

Public Law 91-695, the Lead Based Paint Poisoning Prevention Act (LBPPPA), was signed into law by President Richard Nixon in 1971. The LBPPPA prohibited the use of lead-based paint (defined at that time as 1% lead by weight) in federally assisted housing and gave authority to what was then called the U.S. Department of Health, Education, and Welfare (now the Department of Health and Human Services, DHHS) to make grants to states and local governments for detecting and treating childhood lead poisoning. Further grants were authorized for identifying and abating lead-based paint in public housing. In addition to the DHHS, the U.S. Department of Housing and Urban Development (HUD) was given the authority under the LBPPPA to develop methods of abatement for lead-based paint.

In 1972, HUD published regulations prohibiting the use of lead-based paint in HUD-associated housing. One year later in 1973, amendments were made to the LBPPPA that lowered the lead content allowed in paint to 0.06 percent unless the Consumer Product Safety Commission (CPSC) determined that a higher amount was safe; CPSC subsequently reported that 0.5 percent was a safe level. The 1973 amendments also directed HUD to eliminate, to the extent practicable, all lead paint in HUD-assisted housing constructed before 1950. HUD issued regulations implementing those requirements in 1976.

In 1977 the CPSC, acting under the authority of the Consumer Product Safety Act, declared that paint and similar coatings with a lead content in excess of 0.06 percent lead by weight (dried paint film) are "lead-based paints" and are hazardous materials. CPSC banned the sale of such lead-based paints to consumers and banned the use of lead-based paints where consumers may have direct access to painted surfaces (households, schools, recreation areas, etc.). CPSC also banned products such as toys and furniture that bear lead-based paint.[1] As discussed in Chapter 3, the CPSC 0.06 percent limit still applies to the new purchase of paint but does not apply to paint existing on surfaces.

In 1983 HUD was ordered to restructure its lead-based paint programs by the court in the precedent-setting case known as *Ashton v. Pierce.* At that time, HUD's program considered deteriorating lead-based paint conditions to be hazardous and their remediation efforts were targeted primarily at dwellings of lead-poisoned children. In the court case, public housing tenants alleged that HUD's regulations were deficient for failing to include *intact* lead-based paint as an "immediate hazard," requiring treatment. In 1987, Congress again amended LBPPPA to (1) include intact paint in the definition of "immediate hazard" and establish a new construction cutoff date on 1978; (2) require random sampling for lead-based paint in pre-1978 housing developments; (3) require abatement of lead hazards exceeding 1.0 mg/m^2; and (4) require HUD to conduct extensive research and demonstration programs to assess the cost-effectiveness of alternative abatement strategies.

In 1988, the Stewart B. McKinney Homeless Assistance Amendments Act amended the LBPPPA. Among other things, it (1) eliminated requirements to test and abate public housing units at their turnover and to test 100 percent of all units in a public housing, and (2) established the deadline for lead-based paint testing of public housing authorities to 1994.

THE HUD GUIDELINES

On April 1, 1990, HUD released interim guidelines for public housing authorities to follow in order to identify and abate lead-based paint in all public and Indian housing. Although it established guidelines and requirements applicable only to housing that is owned, subsidized, or subject to mortgage guarantees by the federal government, the significance of the HUD guidelines extends far beyond federally assisted housing. It established "state of the art" procedures for identifying and abating lead-based paint hazards. In the course of their regulatory development, many states and municipalities are considering adoption of various portions of the HUD Guidelines as regulatory requirements to apply to residential buildings.

The guidelines represent the most comprehensive information established to date on testing and abating lead-based paint. Although not mandated for use in private residential or commercial buildings, the guidelines offer a starting point for contractors and building owners in assessing and abating lead-based paint hazards in their buildings. At this writing, HUD is developing more comprehensive guidelines that address risk assessment and interim control measures.

A key provision of the guidelines requires that all housing authorities conduct a random sample of dwelling units (and all common areas) in pre-1978 buildings where children live or are expected to live. This sampling is to be conducted using an X-ray fluorescence analyzer (direct-read or spectrum analyzer). Confirmatory analysis of questionable samples will be verified by atomic absorption spectroscopy analysis (AAS). Under the current guideline requirement, the inspection of all public housing units must be completed by December 1994. Those units with paint containing lead in greater than established action levels will be abated (see Chapter 5 for additional information on testing protocols).

In addition to specifying the number of units to be tested, the guidelines spell out which surfaces are to be tested within each unit. For example, in each area of the unit (an "area" may be a room, closet, hallway, etc.) the guidelines stipulate which components, such as baseboards, ceilings, and chair rails, are to be sampled. The results from this representative testing will be used to draw conclusions, with a high degree of confidence, about the presence of lead-based paint and the need for abatement in all dwellings.

Public housing authorities are required to develop a plan of hazard identification including a final report based on the results of all testing. In the final report, the housing authority must state the location of all lead-based paint, the concentrations of lead found in the paint, and summarize all findings. In addition, the housing authority must institute a periodic surveillance program and develop a progress report that summarizes the progress made to date, problems encountered, and plans for the next reporting period. Should the housing authority change protocols or deviate from the plan in any way, this must be explained in writing.

Once testing is complete, the housing authority must maintain records on which units, common areas, and so on have been tested along with the results of the test. Included with this information is a description of the condition of the painted surface. The authority must provide HUD with information on the identification of the units tested, results of testing, and a description of testing methods used. Finally, the occupants of the unit where lead-based paint is found must be notified of results.

If abatement occurs, the housing authority is obligated to hire contractors who can demonstrate proof of training in the practices and procedures of lead-based paint abatement. Although a nationwide certification and licensing program does not exist at this writing, the guidelines list subjects that must be taught. These include routes of lead exposure, health effects of exposure, worker protection, and specific information on the methodology of abatement to be used in the unit(s). All housing authorities should determine if their state or locality has a licensing and/or certification program in place.

The guidelines provide information on how to find a qualified contractor and what methods of abatement are acceptable. Table 4-1 illustrates a comparison of acceptable abatement strategies, while Tables 4-2 and 4-3 explain appropriate abatement methods to be used on interior and exterior components, respectively. Guidelines issued by HUD also list specific abatement measures that cannot be used in a housing authority. On-site methylene chloride stripping, torch or open-

TABLE 4-1. Comparison of Lead Paint Abatement Strategies

Abatement Strategies	Advantages	Disadvantages
Replacement	Permanent solution; allows for upgrade; can be integrated with modernization; no lead residue left behind on surfaces; low risk of failing to meet clearance standards	Replacement component may be of lesser quality than original; replaced components may be high volume and considered hazardous waste; certain installation requires skill labor
Encapsulation	Low dust if surface preparation is minimal; may be faster than some other methods	May not provide long-term protection; requires routine inspection; may require routine maintenance; quality installation critical for durability
Paint removal on-site	Low level of skill required; allows for restoration	High dust generated; lead residue may remain on substrate and may be difficult to remove; potential difficulty in meeting clearance standards and protecting workers; stripping agents are hazardous and require more precautions.
Paint removal off-site	Allows for restoration; better finished product generally than on-site paint stripping	Lead residue may remain on substrate and may be difficult to remove; damage may occur during removal and reinstallation; swelling of wood, glass breakage and loss of glues and fillers may occur; hardware left on components may be damaged.

Source: HUD Guidelines.

Appropriate Applications	Inappropriate Applications	Comments
Many interior or exterior components; deteriorated components; highly recommended for windows, doors, and easily removed building components	Restoration projects; when historic trust requirements apply; most walls, ceilings, and floors	Nonstandard replacement components may need to be ordered in advance; demolition may damage adjacent surfaces; may result in increased energy efficiency, e.g., window replacement
Exterior trim, walls, floors; interior floors, walls, ceilings, pipes; balustrades, and some interior trim	When encapsulant is not appropriate for substrate and substrate condition; see comments	Must be durable; seams must be sealed to prevent escape of lead dust; safe, effective, and aesthetic encapsulants for interior trim components need to be tested; repainting leaded surfaces and the use of contact paper and paper wall coverings should not be considered for abatement.
Should generally be used on limited surface areas, and when replacement, encapsulation, and off-site removal are impractical; chemical removers work best on metal substrates	Generally should not be done on large surface areas; check with manufacturer regarding recommendations for use on various types of wood and metal substrates	The following are unacceptable methods: • Gas-fired open-flame burning • Grinding or sanding without HEPA filtration • Uncontained water blasting • Open abrasive blasting
Restoration projects, especially doors, mantels, easily removed trim; metal railings	Check with stripping contractor regarding metal substrates	Check with stripping company for timing of work and procedures for neutralizing and washing compounds

TABLE 4-2. Appropriate Abatement Methods by Interior Component

			Remove Paint	
Interior Component	Replace	Encapsulate	On-Site	Off-Site
Cabinets	x			x
Ceilings		x		
Doors	x			x
Floors		x		
Grilles	x			x
Metal pipes		x	x	
Radiators	x		x	x
Soffits		x		
Stair risers		x	x	
Stair treads		x	x	
Trim	x		x	x
balustrades	x	x	x	x
baseboards	x	x	x	x
columns		x	x	
decorative			x	x
door frames	x		x	x
door jambs		x	x	
mantles	x			x
newel posts	x		x	x
plain molding	x	x	x	x
railings	x		x	x
window frames	x		x	x
Walls		x		
Window components				
jambs		x	x	
sashes	x		x	x
sill	x		x	x
stop/parting	x			
well		x	x	
Window units	x			

flame burning, dry abrasive blasting (allowed on exterior abatement only as long as wet methods and high efficiency particle air [HEPA]-filtered vacuums are used to control dust) and machine sanding are all banned.

The HUD guidelines serve as the major source of information for housing authorities who must deal with lead-based paint in their units. In addition, they are an invaluable reference tool for a contractor or consultant entering the lead-based paint testing and/or abatement industry. The "Information Sources" Appendix provides the address and phone number for the HUD USER office, where copies of the HUD guidelines and other pertinent documents may be obtained.

TABLE 4-3. Appropriate Abatement Methods by Exterior Component

Exterior Component	Replace	Encapsulate	Remove Paint On-Site	Remove Paint Off-Site
Ceilings		x		
Doors	x			
Floors		x		
Grilles	x	x		
Metal pipes		x	x	
Soffits		x		
Stair risers		x	x	
Stair treads		x	x	
Trim				
balustrades	x		x	x
columns	x	x	x	
door frames		x	x	x
door jambs		x	x	
newel posts	x		x	x
railings	x		x	x
window frames		x	x	x
Walls		x	x	
Window components				
jambs		x	x	
sashes	x		x	x
sill	x	x	x	
stop/parting	x			
well		x	x	
Window units	x			

TITLE X

The Residential Lead-Based Paint Hazard Reduction Act of 1992 was passed by Congress on October 28, 1992. Commonly called Title X, the law will affect millions of homeowners in the United States. Congress has appropriated $125 million for 1993 and $250 million for 1994 to carry out the provisions of the act.

One of the goals of Title X is lead-based paint hazard assessment and reduction in federally assisted target housing, as opposed to outright abatement of all lead-based paint. "Target housing" is defined as any house built prior to 1978 except housing for the elderly or persons with disabilities (unless a child under 6 lives there). Congress has mandated that a risk assessment must be performed for all federally assisted target housing built before 1978. To assist local and state authorities in promoting these activities in other target housing, HUD has made available funds that can be used for inspections and risk assessments.

By April of 1994, EPA must issue final regulations governing risk assessment, inspection, and abatement procedures, and the training program requirements for each discipline. The Environmental Protection Agency (EPA) will also institute an approval process for training sponsors and delegate that authority to individual states upon an application from the state. However, for any state that does not apply to EPA for approval of their state training and certification program, the law mandates that the state must administer the EPA's program. The EPA will also institute and monitor laboratory certification and publish a list of approved labs annually.

Selling a house will become a bit more complicated under the Title X mandates. By October of 1995, the seller or lessor of target housing must provide the buyer or lessee with a standard lead hazard information pamphlet, which describes the hazards associated with lead-based paint. The seller or lessor must also reveal any information regarding known lead-based paint on the premise and offer the purchaser a 10-day grace period to pull out of the contract, or have the home inspected and a risk assessment conducted.

Once a contract for the sale of the house is drawn, the contract must contain a lead warning statement on a separate piece of paper attached to the contract. By signing the contract, the buyer is acknowledging that he or she has read and understood the lead warning statement, received the lead hazard information pamphlet, and has been allowed at least 10 days to inspect the house for lead-based paint. The lead warning statement must read as follows:

> Every purchaser of any interest in residential real property on which a residential dwelling was built prior to 1978 is notified that such property may present exposure to lead from lead-based paint that may place young children at risk of developing lead poisoning. Lead poisoning in young children may produce permanent neurological damage, including learning disabilities, reduced intelligence quotient, behavioral problems, and impaired memory. The seller of any interest in residential real property is required to provide the buyer of any known lead-based paint hazards from risk assessments or inspections in the seller's possession and notify the buyer of any known lead-based paint hazards. A risk assessment or inspection for possible lead-based paint hazards is recommended prior to purchase.

Numerous exposure studies are mandated by Title X and these include a study of occupational exposure, sources of lead exposure in children, identification of dangerous lead levels, and a study of the exposures received by renovators and remodelers. The EPA will determine if those individuals engaged in renovation and remodeling activities will be subject to the training and certification requirements. Completion of these studies and an evaluation of the results may bring about more or less stringent regulations.

A few of the requirements of Title X have already been implemented. The Occupational Safety and Health Administration (OSHA) has promulgated the interim construction standards for lead, and a public education program has begun. The EPA has a toll-free telephone number for concerned individuals (1-800-LEAD FYI), and a clearinghouse has been instituted to disseminate current knowledge on lead abatement, risk assessment, inspection, and clean-up standards.

Perhaps one of the least known, but most important, aspects of Title X is the requirement that more than one dozen federal agencies—including HUD, EPA, OSHA, CDC, and NIOSH (National Institute for Occupational Safety and Health)—must work together in the development of mandated regulations and guidelines. This requirement is intended to prevent the development of many conflicting regulations among the agencies and allow for easier enforcement of the law. Title X is the first step in achieving that goal and in reducing lead poisoning in America.

The following is a detailed summary of Title X.[2]

Subtitle A—Lead-Based Paint Hazard Reduction

Section 1011—*Grants for Lead-Based Hazard Reduction in Target Housing*—HUD is authorized to provide grants to states or local governmental units to inspect and reduce lead-based paint hazards in target housing. "Target housing" means housing built before 1978 except housing for the elderly or persons with disabilities (unless a child under 6 resides in such housing).

Section 1011(g)—*Technical Assistance and Capacity Building*—HUD is authorized to issue grants of up to $200,000 in fiscal years 1993 and 1994 to establish State training, certification or accreditation programs. The legislation sets aside a total of $3,000,000 in fiscal year 1993 and $3,000,000 in fiscal year 1994 to carry out this subsection. However, section 1011(h) states that grant recipients must contribute at least 10% in matching funds.

Sec. 1011(o)—*Appropriation*—Congress appropriated $125,000,000 for fiscal year 1993 and $250,000,000 for fiscal year 1994 to carry out this Act.

Sec. 1012—*Evaluation and Reduction of Lead-Based Paint Hazards in Federally Assisted Housing*

Sec. 1012(a)—*Requirements for Housing Receiving Fed. Assistance* Note: all the 1012 provisions only apply to housing receiving federal assistance.

Sec. 1012(a)(2)—Legislation amends the Lead-Based Paint Poisoning Prevention Act to remove the mandate that HUD eliminate, as far as practicable, lead paint hazards in intact and nonintact paint that is found in federal housing.

Sec. 1012(a)(3)(A)—*Lead Hazard Information Pamphlet*—Purchasers and tenants of target housing will get a lead hazard information pamphlet that the EPA will develop within 2 years of the Act's enactment.

Sec. 1012(a)(3)(B)—*Risk Assessments*—Target housing constructed prior to 1960 will receive an initial risk assessment by January 1, 1996.

Sec. 1012(a)(3)(B)(i)—(iii) For target housing built between 1960 and 1978,

25% of the housing will receive an initial risk assessment by January 1, 1998; 50% by January 1, 2000, and the rest by January 1, 2002.

Sec. 1012(a)(3)(C)—HUD procedures will provide for inspections when federally funded renovation or rehabilitation projects are likely to disturb painted surfaces.

Sec. 1012(a)(3)(D)—HUD will have procedures to reduce lead paint hazards in rehab. projects receiving less than $25,000 per unit in federal funds.

Sec. 1012(a)(3)(E)—HUD procedures will provide for abatement of lead-based paint hazards when rehab. projects receive more than $25,000 per unit in federal funds.

Sec. 1012(a)(3)(F)—where risk assessment, inspection, or reduction activities occur, occupants will receive notice of nature and scope of activities, and the actual risk assessment or inspection reports.

Sec. 1012(g)—*Section 8 Housing*—HUD has the discretion to raise rents in Section 8 rental housing to cover the lead inspection and abatement costs. (Amends Section 8 of the U.S. Housing Act of 1937).

Sec. 1012(h), (i), and (j)—*HOPE Programs*—Planning grants under three federal programs—the HOPE for Public and Indian Housing Homeownership program, the HOPE for Homeownership of Multifamily Unit program, and the HOPE for Single Family Homes program—may be used to inspect for lead-based paint hazards. Similarly, implementation grants under the same three programs may be used to abate lead-based paint hazards. (Amends sections of the U.S. Housing Act of 1937).

Sec. 1012(k) and (l)—*FHA Insurance*—As part of FHA insurance for single family homes and multifamily housing, HUD is able to provide home improvement loans that allow for the "evaluation and reduction of lead-based paint hazards. Rehabilitation loans for 1 to 4 family structures may provide for measures to evaluate and reduce lead-based paint hazards. (Amends section 2(a) and 203(k)(2)(B) of the National Housing Act.)

Sec. 1012(m)—*Rural Housing*—The Secretary of Agriculture can give financial assistance to farmers and homeowners in rural areas to repair and rehab their dwellings which may provide for measures to evaluate and reduce lead-based paint hazards. (Amends section 501(a) of the Housing Act of 1949).

Sec. 1013—*Disposition of Federally Owned Housing*—The section deletes the following sentence from the Lead-Based Paint Poisoning Prevention Act, "Further, the Secretary shall establish and implement procedures to eliminate the hazards of lead based paint poisoning in all federally owned properties prior to the sale of such properties when their use is intended for residential habitation." Instead, this section requires the following:

(A) *For Pre-1960 Federally-Owned Target Housing*—Beginning on January 1, 1995, HUD will have procedures to inspect and abate lead-based paint hazards in all federally owned target housing built before 1960.

(B) *For Federally-Owned Target Housing Built Between 1960 and 1978*—Beginning on January 1, 1995, HUD will have procedures to inspect for lead-based paint and lead-based paint hazards in all federally owned target housing built between 1960 and 1978. The inspection results "shall be made available" to prospective purchasers. HUD can waive the requirements of this subsection if a certified contractor determined no lead-based paint hazards are present.

(C) *Budget Authority*—The requirements of (A) and (B) apply only if appropriations are available. If paragraphs (A) and (B) impose additional costs to the Resolution Trust Corp. or the Federal Deposit Insurance Corp., its requirements apply only if an appropriation is provided in advance in an appropriations act.

Sec. 1014—*Comprehensive Housing Affordability Strategy*—To be eligible for assistance under the National Affordable Housing Act, a State or local governmental unit must submit a housing strategy that, among other things, estimates the number of low-income homes that contain lead-based paint hazards, outlines the actions proposed or being taken to reduce lead-based paint hazards, and describes how lead-based paint hazard reduction will be integrated into housing policies and programs. (Amends section 105 of the National Affordable Housing Act.)

Sec. 1015—*Task Force on Lead-Based Paint Hazard Reduction and Financing*—HUD, in consultation with the EPA, will establish a task force to recommend ways to expand resources and efforts to evaluate and reduce lead-based paint hazards in private housing. The Task Force's responsibilities include, among other things, recommendations to incorporate lead hazard concerns into underwriting standards, appraisal guidelines, and home inspections for mortgages, and clarifying landlord and lender liability.

Sec. 1016—*National Consultation*—In carrying out this Act, HUD must consult with EPA, CDC, other concerned federal agencies, and the task force established in section 1015.

Sec. 1017—*Guidelines for Evaluation and Reduction Activities*—Within 12 months of the date of the Act's enactment, HUD, in consultation with EPA, the Secretary of Labor, and the CDC, shall issue guidelines for the conduct of federally supported work involving risk assessment, inspections, interim control and abatement. Guidelines will be based upon criteria that measure housing condition and the presence of children under age 6.

Sec. 1018—*Disclosure of Information Upon Purchase, Sale or Lease of Target Housing*

Sec. 1018(a)(1)—Within 2 years of the Act's enactment, HUD and EPA shall issue regulations that require that before the purchaser or lessee is obligated under any contract, the seller or lessor shall:

(A) Provide a lead hazard information pamphlet, as prescribed by the EPA
(B) Disclose any known lead-based paint or any known lead-based paint hazard

and to provide any lead hazard evaluation report available to the seller or lessor or lessor

(C) Permit the purchaser a 10 day period (or, upon mutual agreement, a different period of time) to conduct a risk assessment or inspection for lead-based paint hazards.

Sec. 1018(a)(2)—*Contract for Purchase and Sale*—The regulations will require that every contract to purchase and sell target housing will have a lead warning statement and a statement signed by the purchased that he or she read and understood the lead warning statement, received a lead hazard information pamphlet, and had a 10 day opportunity, or other agreed upon period of time to inspect for lead-based paint hazards.

Sec. 1018(a)(3)—*Contents of Lead Warning Statement*—In large type on a separate piece of paper attached to the contract, the Lead Warning Statement will provide that:

> Every purchaser of any interest in residential real property on which a residential dwelling was built prior to 1978 is notified that such property may present exposure to lead from lead-based paint that may place young children at risk of developing lead poisoning. Lead poisoning in young children may produce permanent neurological damage, including learning disabilities, reduced intelligence quotient, behavioral problems, and impaired memory. Lead poisoning also poses a particular risk to pregnant women. The seller of any interest in residential real property is required to provide the buyer with any information on lead-based paint hazards from risk assessments or inspections in the seller's possession and notify the buyer of any known lead-based paint hazards. A risk assessment or inspection for possible lead based paint hazards is recommended prior to purchase.

Sec. 1018(a)(4)—*Real Estate Agent Liability*—Real estate agents have an obligation to ensure compliance with this section.

Sec. 1018(a)(5)—*Citizen Suits*—Citizens may sue HUD or EPA to compel issuance of regulations under this section.

Sec. 1018(b)—*Penalties*—A knowing violation of this section subjects a person to civil penalties in the amount triple the damages incurred by the purchaser or lessee, reasonable attorney fees, expert witness fees, and costs. A failure or refusal to comply with this section also subjects a person to enforcement action under the Toxic Substances Control Act and penalties up to $10,000 per violation.

Sec. 1018(c)—*Validity of Contracts and Loans*—Nothing in this section affects the validity or enforceability of any sale or contract for sale or any loan, mortgage or lien.

Sec. 1018(d)—*Effective Date*—The regulations under this section will take effect 3 years after the Act's date of enactment.

Subtitle B—Lead Exposure Reduction

Sec. 1021—Amends the Toxic Substances Control Act by adding a Title IV entitled "Lead Exposure Reduction"

Sec. 402—*Lead-Based Paint Activities Training and Certification*

Sec. 402(a) (1)—*Regulations*—Within 18 months of the Act's enactment, EPA will issue final regulations governing risk assessment, inspection, and abatement "to ensure that individuals engaged in such activities are properly trained; that training programs are accredited; and that contractors engaged in such activities are certified."

Sec. 402(a) (2)—*Accreditation of Training Programs*—EPA regulations will include requirements for the accreditation of risk assessment, inspection, and abatement training programs for workers, supervisors, inspectors and planners, and other individuals involved in lead-based paint activities.

Sec. 402(a) (3)—*Accreditation and Certification Fees*—EPA (or a State with an authorized program) shall impose a fee on those who operate training programs and on certified contractors. The fee will not be imposed on any governmental or nonprofit training program. The fee may also be waived for a contractor who trains his own employees.

Sec. 402(c) (1)—*Renovation and Remodeling*—Within 18 months of the Act's enactment, EPA will issue guidelines for renovation and remodeling work that may create an exposure risk to dangerous levels of lead. The guidelines will be distributed through hardware and paint stores, employee organizations and trade groups, and State and local agencies.

Sec. 402(c) (2)—*Study of Renovators' and Remodelers' Exposure and Creation of a Hazard*—Within 30 months of the Act's enactment, EPA will conduct a study of lead exposure among renovators and remodelers and whether their activities create a lead-based paint hazard.

Sec. 402(c) (3)—*Certification Determination*—Within 4 years of the Act's enactment, EPA will apply the training and certification requirements to renovation and remodeling contractors unless the EPA study demonstrates that a category of contractors engaged in renovation or remodeling does not require certification.

Sec. 403—*Identification of Dangerous Lead Levels*—Within 18 months of the Act's enactment, EPA will issue regulation that identify lead-based paint hazards, lead-contaminated dust, and lead-contaminated soil.

Sec. 404—*Authorized State Programs*

Sec. 404(a)—*Approval*—A State training and certification program and/or a State program to distribute lead hazard information pamphlets may be authorized to operate in lieu of the federal program if the State certifies in an application that (1) the State program is at least as protective of human health and the environment as the federal program; and (2) the State program provides adequate enforcement.

Sec. 404(b)—*Approval or Disapproval*—The EPA has 180 days following submission of an application to approve or disapprove the application. EPA may approve State program only if it finds that the program meets the requirements outlined in (1) and (2) above.

Sec. 404(c)—*Withdrawal of Authorization*—If a State program is disapproved, it has 180 days to correct the deficiencies or the Administrator shall withdraw authorization and establish a Federal program.

Sec. 404(d)—*Model State Program*—Within 18 months after the Act's enactment, EPA shall issue a model State program which a State may adopt to administer and enforce as an authorized State program.

Sec. 404(e)—*Other State Requirements*—A State any impose requirements more stringent than this Act.

Sec. 404(f)—*State and Local Certification*—A State or local government may not require a contractor to obtain more than one certification to carry out lead-based paint activities.

Sec. 404(g)—*Grants to States*—EPA may issue grants for the States to implement State programs.

Sec. 404(h)—*EPA Enforcement*—A State must have an authorized State program on training and certification and on the lead hazard information pamphlet within 2 years of the EPA regulations of the State will be required to administer EPA's federal program.

Sec. 405—*Lead Abatement and Measurement*

Sec. 405(a)—*Program to Promote Abatement*—EPA will conduct a comprehensive program on safe, effective, and affordable monitoring, detection and abatement of lead.

Sec. 405(b) (1)—*Standards for Environmental Sampling Labs*—Within 2 years of the Act's enactment, EPA, in consultation with the Dept. of Health and Human Services will establish a program to certify labs to test substances for lead unless EPA determines that effective voluntary accreditation programs are in place on a nationwide basis.

Sec. 405(b) (2)—*List of Environmental Labs*—Within 2 years of the Act's enactment and annually thereafter, EPA will publish a list of certified or accredited environmental labs.

Sec. 405(b) (3)—*Review of Voluntary Accreditation*—If the EPA determines that voluntary accreditation is sufficient, the EPA will review its determination after 3 years, and if the voluntary program is not effective, the EPA will establish a certification program within the following year.

Sec. 405(c)—*Exposure Studies*

Sec. 405(c) (1)—*Sources of Lead Exposure for Children*—CDC and the National Institute of Environmental Health Sciences will jointly study sources of lead exposure in children who have elevated blood lead levels and provide a copy of the report to Congress within 30 months of the Act's enactment.

Sec. 405(c) (2)—*Study of Occupational Exposure*—HUD, in consultation with the National Institute for Occupational Safety and Health, will study means to reduce hazardous occupational lead abatement exposure and provide a copy of the report to Congress within 30 months of the Act's enactment.

Sec. 405(d)—*Public Education*—EPA, HHS, and HUD will sponsor public education and outreach activities on lead poisoning, the need for universal screening, risk assessment and inspection methods, the need for quality abatement, and measures to reduce lead exposure. EPA will also develop information to be distributed to retailers of home improvement products to warn consumers of the hazards of renovation and remodeling where lead-based paint may be present.

Sec. 405(e) (1)—*Clearinghouse*—Within 6 months of the Act's enactment, EPA, in consultation with HUD and CDC, will establish a National Clearinghouse on Childhood Lead Poisoning to collect, evaluate, and disseminate information on the assessment and reduction of lead-based paint hazards, detection and risk assessment methods, abatement and clean-up standards.

Sec. 405(e) (1) (b) and (e) (2)—*Rapid-Alert System and Hotline*—Within 6 months of the Act's enactment, the EPA will have a rapid-alert system to inform certified contractors of significant developments and will establish a hotline to provide the public with answers to lead poisoning prevention questions.

Sec. 405(f)—*Products for Lead-Based Paint Activities*—Within 30 months of the Act's enactment, the President shall establish appropriate criteria, testing protocols and performance standards to ensure that lead hazard evaluation and reduction products are effective.

Sec. 406(a)—*Lead Hazard Information Pamphlet*—Within 2 years of the Act's enactment, EPA, in consultation with HUD, and HHS, shall publish a lead hazard information pamphlet that contains information on the health risks associated with lead exposure, provides information on the presence of lead-based paint hazards, describes the risks of renovation in a dwelling with lead-based paint hazards; provides information on approved methods to evaluate and reduce lead-based paint hazards, recommends a risk assessment or inspection prior to the purchase, lease, or renovation of target housing, advises persons how to obtain a list of certified contractors, and provides a listing of Federal, State and local agencies that can provide information about applicable laws.

Sec. 406(b)—*Renovation of Target Housing*—Within 2 years of the Act's enactment, EPA shall issue regulations that requires paid renovators of target housing to provide a lead hazard information pamphlet to the owner and occupant of such housing before starting the renovation.

Sec. 408—*Control of Lead-Based Paint Hazard at Federal Facilities*—Federal agencies must comply with all Federal, State, interstate and local requirements relating to lead-based paint activities and lead-based paint hazards, and the United States waives any immunity with respect to these requirements. However, no federal employee is personally liable for any civil penalties for any act or omission within the scope of his official duties.

Sec. 409—*Prohibited Acts*—It is unlawful to fail or refuse to comply with any provision of this title or any rule or order issued under this title.

Sec. 410—*Other Federal Laws*—Other federal agencies may establish or enforce any requirements which are at least as stringent as those established pursuant to this title.

Sec. 411—*Administrative Proceedings*—EPA will establish a rulemaking docket for actions taken pursuant to this act and will allow interested parties to inspect and copy the docket. Transcripts of public hearings and written comments on proposed rules will be included in the docket.

Sec. 412—*Specific Enforcement*—United States district courts will have jurisdiction to compel action or restrain any violation under this title.

Subtitle C—Worker Protection

Sec. 1031—*Worker Protection*—Within 180 days of the Act's enactment, the Secretary of Labor shall issue an interim final regulation regulating occupational exposure to lead in the construction industry. These regulations will provide employees with as safe a workplace as would prevail under HUD guidelines and will have the legal effect of an Occupational Safety and Health Standard.

Sec. 1033—*NIOSH Responsibilities*

1. *Training Grants*—The Occupational Safety and Health Act of 1970 is amended to authorize the National Institute for Occupational Safety and Health (NIOSH) and EPA to make grants to nonprofit organization, States, and nonprofit government employee organizations to train and educate workers and supervisors who are or may be directly engaged in lead-based paint activities. The Institute will receive at least 10 million dollars in each fiscal year from 1994 through 1997 to make grants under this section.
2. *Evaluation of Training Programs*—NIOSH is given $500,000 in each fiscal year 1994 through 1997 to evaluate the training programs that received grants and to include recommendations when appropriate to revise these programs.

Subtitle D—Research and Development

Part 1—HUD Research

Sec. 1051—*Research on Other Sources*—HUD will conduct research on strategies to reduce lead exposure from other sources including exterior oil and interior lead dust in carpets, furniture, and forced air ducts.

Sec. 1052—*Testing Technologies*—HUD will conduct research to

1. Develop methods to evaluate lead-based paint hazards in housing
2. Develop methods to reduce lead-based paint hazards in housing

3. Develop improved methods to measure lead in paint, films, dust, and soil
4. Establish performance standards for detection methods, including spot test kits
5. Establish performance standards for hazard reduction methods, including encapsulants
6. Establish appropriate cleanup standards
7. Evaluate the efficacy of interim controls
8. Evaluate the relative performance of abatement techniques
9. Evaluate long-term cost-effectiveness of interim control and abatement strategies
10. Assess the effectiveness of hazard evaluation and reduction activities funded by this Act

Sec. 1053—*Authorization*—Of the total amount approved in appropriations under sec. 1011(o), HUD may use $5 million in 1993 and $5 million in 1994 to conduct the research in this part.

Part 2—GAO REPORT

Sec. 1056(a)—*Federal Implementation Study*—The US Comptroller General will assess the effectiveness of federal enforcement and compliance with lead safety laws and regulations, including any changes needed in annual inspection procedures to identify lead-based paint hazards in Section 8 housing.

Sec. 1056(b)—*Insurance Study*—The Comptroller General will assess availability of liability insurance and recommend measures to increase the availability of this insurance for owners of residential housing and persons engaged in lead-based paint evaluation and reduction activities.

Subtitle E—Reports

Sec. 1061(a)—*Annual HUD Reports*—Each year, HUD will transmit a report to Congress that assess the programs authorized by this law, that summarizes studies on childhood lead poisoning, and recommends legislative and administrative initiatives that would better enable HUD to combat lead hazards.

Sec. 1061(b)—*Biennial HUD Reports*—Beginning 2 years after this Act's enactment and every 2 years thereafter, HUD will report to Congress on its implementation and effectiveness of lead-based paint hazard evaluation and reduction activities.

Notes

1. 42 *Federal Register* 4419, September 1, 1977.
2. April 1993 Title X summary prepared by Susan Silver, Director of the Office of Inmate Advocacy, New Jersey Department of the Public Advocate.

5

Overview of Lead-based Paint Testing Methods

Kenneth T. White, M.S., M.B.A. (M.M.), C.I.H., C.S.P.

Vincent M. Coluccio, Dr.P.H.

FUNDAMENTALS

Units of Measurement

Levels of lead in dried paint films are measured in units of either mass of lead per mass of sample or mass of lead per unit area of the surface sampled.

To determine mass of lead per mass of sample, a bulk sample of paint chips or paint scrapings is analyzed in a laboratory by atomic absorption spectroscopy (AAS) or other comparable analytical techniques. The analytical instrument provides results in units of mass concentration per volume of solution prepared from a known-mass portion of the paint sample. The laboratory then mathematically converts to units of percent lead by weight (percentage of lead per total mass of the sample) or parts per million of lead (ppm, parts of lead per million parts of sample), for example:

100% lead by weight	=	1,000,000 ppm lead
4% lead by weight	=	40,000 ppm lead
0.5% lead by weight	=	5,000 ppm lead

From these examples, we can see that multiplying units of percent lead by weight by a factor of 10,000 yields an equivalent measure in units of ppm. Conversely, dividing units of ppm by 10,000 provides an equivalent measure in percent lead by weight.

Determination of mass of lead per unit area is typically accomplished by use of hand-held X-ray fluorescence (XRF) analyzers, which directly provide data in units of milligrams of lead per square centimeter of the sampled surface area (mg/

cm^2). These same units of measurement can also be obtained by laboratory analysis of paint chips or scrapings taken from a known surface area, read as total milligrams of lead in the sample per the measured area from which the sample was taken.

Regulatory Definitions of Lead-Based Paint

The Residential Lead-Based Paint Hazard Reduction Act of 1992 (commonly called Title X) among other things defines lead-based paint and amends the Toxic Substances Control Act (TSCA). Title X defines paint as lead-based if it a has dry film lead mass concentration equal to or greater than 0.5 percent lead by weight (5,000 ppm) or a dry film area concentration greater than or equal to 1.0 mg/cm^2. Title X also defines lead-based paint as a toxic substance, under TSCA. As lead-based paint is a toxic substance, other housing regulations require federal housing authorities to abate any and all lead-based paint in federally owned or subsidized dwellings where children live or could be expected to live.

States and localities use these (0.5% and 1.0 mg/cm^2) or similar values to trigger abatement of painted surfaces that are deteriorated or directly accessible to young children. Actual numerical criteria for classifying paint as lead-based vary from state to state, and even from one municipality to another within a state. For instance, the Massachusetts definition for lead-based paint is 1.2 mg/cm^2. In both New York City and Baltimore, lead-based paint is defined as 0.7 mg/cm^2. But, in all three locations, lead-based paint, as expressed in units of mass of lead per mass of sample, is 0.5 percent by weight lead (5,000 ppm). In New Jersey, however, the limits are 1.0 mg/cm2 and 1.0 percent lead by weight.

In light of this diversity of definitions, it is advisable to contact lead poisoning prevention programs to obtain the precise definition used in the particular jurisdiction within which you will be working. Where state and local regulations conflict with federal regulations, lead-based paint inspection and abatement activities must comply with the most stringent requirements. Appendix C provides a listing of federal and state lead poisoning prevention programs.

Prior to government regulations, some interior residential paints contained large amounts (sometimes greater than 50%) of lead. The U.S. Consumer Product Safety Commission (CPSC) in 1978 banned the sale of lead-based paint to consumers and banned the use of lead-based paint in residences and other areas where consumers have direct access to painted surfaces. At the time of the CPSC ban, lead-based paints were defined as paints containing more than 0.06 percent lead by weight (600 ppm lead) taken from dry-to-the-touch, newly applied paint films.

The 1978 ban is still in effect and it is important to keep in mind that it prohibits further sale of residential paints that would create dry paint films with lead levels greater than 0.06 percent. Stated differently, the CPSC limit effectively applies to paint "in the paint can" and prohibits the use of that paint if its use results in dry paint films with greater than 0.06 percent lead. The CPSC has no mandate to address the leaded paint already in housing.

In some cases the 0.06 percent lead definition has been suggested as a trigger for abating paint. This practice would surely lead to widespread abatement, including quite possibly some paints with low and perhaps nonhazardous concentrations of lead.

Limitations of Lead-Based Paint Definitions

The different definitions, and the application of those different definitions, of lead-based paint as discussed above need to be understood by practitioners engaged in lead-based paint testing and abatement activities.

The levels that define lead-based paint were set partly because they represented the lowest that could reliably be measured by field testing in situ. When dealing with lead, even low levels may be toxic. In fact, there is no reliable and consistent connection between the concentration of lead in a paint film and the risk of adverse health effects. That connection would have to be established by relating the lead concentration to deterioration or activities that cause release of lead particulates from the paint film, and consequent human exposures to and uptake of those particulates.

For example, a predictive modeling study performed by the State of California calculated the airborne lead levels that would arise when alternative abatement techniques were performed on a surface with a paint film containing 1.0 mg/cm^2 of lead.[1] Wet scraping of the surface was calculated to result in 37 micrograms of lead per cubic meter of air ($\mu g/m^3$); however, dry scraping of the same surface was estimated to result in an exposure level of about 371 $\mu g/m^3$, fully a tenfold increase in predicted airborne lead concentrations.

Based on the lack of a reliable connection between lead-in-paint concentrations and airborne lead levels, worker safety and health provisions of the U.S. Occupational Safety and Health Administration's (OSHA) new standard for protecting construction workers from lead exposures are triggered by airborne lead levels, which in turn are based on health effects data.[2] Worker protections under the OSHA standard are not triggered by a specific lead-in-paint concentration, since in OSHA's words, "OSHA would have no idea what health effects might be triggered by such a surface concentration." The standard used by OSHA, the agency states, ". . . applies to all occupational exposures to lead in all construction work in which lead, *in any amount,* is present in an occupationally related context" (emphasis added).

Thus, OSHA requires that precautionary measures in the form of respiratory protection and airborne lead monitoring be conducted when workers disturb paint with any amount of lead in it. Worker protections then are relaxed or tightened based on the outcome of lead dust in-air testing.

Unlike OSHA's health-based approach to protecting workers, current regulations do not mandate occupant protections when renovation or demolition is performed involving lead-containing paints with lead concentrations below the definition of lead-based paint. In the case of a federally subsidized public housing project for example, young children and other occupants may be present and unprotected from lead-in-air exposures while paint of 0.4 percent or 0.8 mg/cm^2 lead (less than the 0.5% or 1.0 mg/cm^2 Title X level) is being stripped in their apartment, despite lack of knowledge of the airborne and surface lead dust levels created by the stripping. In this scenario, however, the workers doing the paint stripping are covered under the OSHA lead construction standard and therefore their exposures to airborne lead must be monitored, and, if it exceeds OSHA's limit, the workers must be protected.

Based on the foregoing discussion, lead-based paint inspectors and contractors cannot rely on the regulatory definitions of lead-based paint as the threshold to determine hazard or no-hazard. When testing indicates measurable lead-in-paint concentrations, even if below the limits for regulation-defined lead-based paint, it is advisable to reference the test findings in the context of applicable regulations and limits but not to state that the surface is "safe" or that abatement may commence without occupant protections. Indeed, prudence seems to dictate that abatement precautions be taken when *any* level of lead is identified in a paint film.

Another significant limitation in the definitions of lead-based paint is that units of mass of lead per sample mass (percentage lead or ppm) do not correspond to units of mass of lead per unit area (mg/cm^2). Indeed, the use of laboratory analysis of paint chip samples and XRF testing of the same paint film may provide two different outcomes. Consider the following scenarios encountered in a federally subsidized public housing authority:

Scenario A: Wall "A" is coated with one thin layer of paint resulting in a paint film of 0.9 percent lead by weight. Laboratory testing of a scrape sample from this wall will indicate that lead is present at 0.9 percent lead by weight, a level that is greater than HUD's limit of 0.5 percent and therefore must be considered by HUD to be a lead-based paint. However, an XRF test of the thin paint film on this wall did not detect lead in levels greater than HUD's alternative area concentration limit of 1.0 mg/cm^2. The "negative" findings of the XRF test will therefore conflict with the "positive" results of the laboratory analysis.

Scenario B: Wall "B" is coated with many layers of paint, each layer containing 0.3 percent lead by weight. Laboratory testing of a scrape sample from this wall indicated that lead is present at 0.3 percent lead by weight, a level that is less than HUD's limit of 0.5 percent and therefore is considered by HUD not to be lead-based paint. However, an XRF test of the multiple layers of paint on this wall detected lead in levels greater than HUD's alternative limit of 1.0 mg/cm^2. The "positive" findings of the XRF test therefore conflict with the results of the laboratory analysis.

As the preceding scenarios illustrate, the selection of a testing method can influence the test outcome. By reporting all lead values in the same units of measurement for a given test site, the problem of conflicting outcomes can be avoided. For example, when paint scrape samples are analyzed by a laboratory to confirm XRF measurements, the laboratory should be directed to report results in the same units of measurement as the XRF (i.e., mg/cm^2) and not in percent or ppm lead (to assure that this is possible, a known area of the surface must be carefully measured and scraped to create the sample for the lab to analyze).

OVERVIEW OF TESTING AND MONITORING METHODS

All lead abatement projects require testing and/or monitoring. Many of these procedures and protocols are ill defined in the minds of those involved, and some do not see the depth and breadth of the program. Testing and monitoring is an integral part of the lead abatement process and provides the quantitative measures by which the success of abatement is determined.

To begin the discussion, tests for lead content can be divided into classes to explain both how they are performed and to give comparative information on attributes. Tests can be divided into three categories: analytical tests, survey tests, and screening tests.

Analytical Tests

For this discussion, analytical tests are defined as those that:

- Have a very high level of element specificity. This means that the test is able to determine the presence and concentration of lead in the sample, no matter what else is present.
- Have results that are highly accurate. They provide data that are correct. They provide the "right" answers using statistically based quality control processes to monitor and assure accurate data generation. The results of this program become part of the record of the data reported to the client.
- Have results that are highly precise. They provide data that are reproducible. Again using statistically based processes, they provide a high degree of certainty that if another analysis of the sample is performed, the statistically same data would result. The results of this program also become part of the record of the data reported to the client.
- Have results generated using a documented process whose procedures are followed sample-to-sample, day-to-day, week-to-week, and so on. These are called standard operating procedures (SOPs). These SOPs are built from either nationally accepted procedures or are procedures that have been validated. The use of these SOPs assures that the proper procedures are performed and that

they are performed correctly. Key to this is the requirement that the SOPs be fully written down and readily available and used by those personnel assigned to perform them. Further, any changes or modifications to the SOP (required by some unique situation) should be recorded in detail to become part of the record.

- Have a fully documented and executed procedure for instrument calibration. The procedure incorporates the use of nationally recognized standards (traceable back to the National Institute of Standards and Technology, NIST) and is conducted over the entire range of expected results. Calibration first qualifies use of the instrument by conduct of a procedure to determine if the instrument is functioning within defined parameters and then continues as long as data is generated. The initial calibration procedure includes use of blanks and at least three knowns. The range of these three or more knowns defines the range within which data may be reported (i.e., report no data smaller than the smallest known nor larger than the largest known). Calibration is conducted as the instrument is in use. This process notifies the operator when adjustments to the instrument need be performed to regain optimal instrument performance. As before, procedures are written, available for use, and used routinely by operators. Any modifications are documented to the record.

The use of analytical tests, practically speaking, means taking a sample (intrusive sampling) and sending it to a laboratory for analysis, where at least a portion of the sample is consumed in analysis (destructive testing). The laboratory has the clean, secure environment needed to conduct the extensive quality assurance processes required by analytical test procedures. Again speaking practically, conduct of these procedures is at best difficult, and conduct in environments other than those that are tightly controlled, as in a laboratory, are quite suspect.

Instruments used to conduct analytical testing for lead are listed below. The list is not all-inclusive but gives those most commonly found.

- *Atomic absorption (AA) and graphite furnace atomic absorption (GFAA) spectrometer.* Most commonly used as it is the least expensive of the popular instrument types and is more easy to use than most. As it is the most commonly used instrument type, there are many nationally recognized methods for its use in analysis for lead in various matrices. Individual instruments may use a flame (AA) or a furnace (GFAA) as the excitation source used to perform the analysis.
- *Inductively coupled plasma spectrometer (ICP).* In extensive use as it is faster than AA/GFAA; some can perform analyses for multiple metals simultaneously. Usually uses an argon plasma as excitation-energy source. Measures emitted light in atomic emission spectroscope version (ICP-AES), or separates and quantifies individual atomic isotopes in mass spectrometer version (ICP-MS).

- *X-Ray fluorescence analyzer (XRF).* Available as a laboratory analytical instrument, but in less common use. Detects and measures emitted X-rays from samples that are being irradiated with gamma radiation. Uses a radioactive isotope as excitation source.
- *Ultraviolet/visible spectrometer.* Available, but in less common use. Measures the light absorbed by lead solutions at a specific frequency/wavelength (spectroscopic analysis). Compounds that form colored solutions in the presence of lead may also be used and the light absorbed is then measured at a specific frequency/wavelength as before (colorimetric analysis).
- *Voltametric cell.* Blood analysis method in common use utilizes anodic stripping voltametry (ASV) to deposit metals in blood onto an electrode and then selectively and quantitatively strip-off the lead. Not commonly in use for other matrices.

Because of their high degree of documented statistically based quality, data generated by analytical tests conducted in a qualified third-party laboratory are considered the "gold standard" for data and are admissable as evidence in a court of law. The data generated are also usually the most expensive.

Survey Tests

Survey testing for lead is conducted in the field (hence "survey") and yields a numerical result (a number, a value). Measurement of lead content in paint may require a survey that is nondestructive, using an instrument that directly measures lead concentration, or destructive, requiring collection of a sample for on-the-scene analysis. (see Figs. 5-1 and 5-2.)

Methods used to conduct survey testing involve various measurement techniques. These include rare but available use of ASV and colorimetric instruments. Also included is use of a simple visual-comparison colorimetric method performed in a manner similar to the determination of chlorine in swimming-pool water. But by far the most common lead survey testing instrument is the field-portable XRF.

The XRF uses a radioactive source (gamma emitter) to ionize lead atoms in a paint sample. These lead ions (ionized lead atoms) return to a normal state and emit X-rays that are characteristic of lead. The XRF detects and quantifies these emitted X-rays to determine the amount of lead present.

The XRF displays the lead concentration reading in units of milligrams of lead per square centimeter of surface area (mg/cm^2). The limits of detection and maximum quantifiable amounts of lead vary among the available instruments. For instance, one analyzer will indicate a maximum of 10 mg/cm^2 while another instrument has the ability to quantify up to 100 mg/cm^2. When surveying a site for

NATIONAL LEAD INFORMATION CENTER HOTLINE, 1-800-LEAD-FYI

TESTING YOUR HOME FOR LEAD

WHO SHOULD DO THE TESTING?

Testing to determine the presence of lead in paint, dust, water, and soil is best done by trained professionals. Contact your local public health organization for information on lead inspection services and testing laboratories in your area. Other community organizations may be able to provide information or referrals for inspection services.

Some states require that lead inspectors be state-certified. If your state has this requirement, be sure to hire a state-certified inspector.

TESTING FOR LEAD IN PAINT

To thoroughly analyze the paint in your home, each different painted surface should be tested. Different paints may have been used on walls, window frames, doors, and so on. Paints may also differ from room to room. Each of your home's painted surfaces, both inside and outside, should be tested separately.

Professional testing companies use two basic methods to measure lead in paint:

- X-ray fluorescence (XRF) uses portable detectors that X-ray a painted surface to measure the amount of lead in all the layers of paint. This type of testing is done in the home and disturbs little, if any, paint.
- Laboratory testing of paint samples involves removing samples of paint from each surface to be tested, usually from an area of about two square inches. Samples are sent to laboratories for analysis. This method leaves a bare spot on each surface tested.

Do-it-yourself test kits are commercially available. These kits do not tell you how much lead is present, however, and their reliability at detecting low levels of lead has not been determined. Professional testing for lead in paint is recommended over the do-it-yourself test kits. (See "Home Test Kits for Lead in Paint, Soil and Dust" on the reverse side of this sheet for additional information about home test kits.)

TESTING FOR LEAD IN HOUSEHOLD DUST

Household dust may contain tiny particles of lead released from lead-painted surfaces inside the home or tracked in with lead-contaminated soil from outside.

The recommended sampling method for dust is the surface wet wipe. Dust samples are collected from different surfaces, such as bare floors, window sills, and window wells. Each sample is collected from a measured surface area using a wet wipe, which is sent to a laboratory for testing.

TESTING FOR LEAD IN WATER

Household drinking water may contain lead, usually from old pipes that contain lead or from lead solder in plumbing.

Water samples can be collected directly from the faucet. Two samples are usually collected. One sample is taken from water that has been standing in the plumbing lines overnight or for eight hours or longer, and a second sample is taken after letting the water run for several minutes to flush the lines. The water samples are then sent to a laboratory for analysis.

TESTING FOR LEAD IN SOIL

Lead may be present in the soil around your home and near streets and highways close to your home.

To test soil for lead, samples are taken from areas near your home where children play and from areas that contain soil likely to be tracked into your home. The soil samples are sent to laboratories for analysis.

"Testing Your Home for Lead" was prepared by the U.S. Army Construction Engineering Research Laboratories, Champaign, Illinois, for distribution by the National Lead Information Center. 6/93

The National Lead Information Center is operated by the National Safety Council with funding from the Environmental Protection Agency, the Centers for Disease Control and Prevention, the Department of Housing and Urban Development, and the Department of Defense.

FIGURE 5-1. National Lead Information Center fact sheet concerning testing homes for lead.

NATIONAL LEAD INFORMATION CENTER HOTLINE, 1-800-LEAD-FYI

HOME TEST KITS FOR LEAD IN PAINT, SOIL, AND DUST

What are home test kits?

Home test kits are chemical test kits designed to check for lead in paint, soil, and dust. These kits, which can be used by consumers or professional inspectors, are relatively inexpensive, provide quick results, and are widely available at retail hardware and home improvement stores.

How do home test kits work?

There are a number of different test kits on the market, and procedures for their use vary. Most test kits contain a chemical (either rhodizonate or sodium sulfide) that changes color in the presence of lead.

My home was built before 1978, so it may contain lead-based paint. Should I use a test kit to see whether my home has a lead problem that should be abated?

Though home test kits show promise for the future, their accuracy has not yet been proven by independent field study. For this reason, federal agencies do not currently recommend using test kit results as the basis for making decisions about abatement of lead in paint, soil, or dust. This applies to all home test kit results, including home test kit results obtained by professional inspectors.

Is the federal government currently doing any research on home test kits?

Yes. Several test kit evaluations are now under way. After those evaluations are completed in late 1993 and early 1994, updated information will be made available through the National Lead Information Center.

How can I get the paint in my home tested for lead?

Professional testing for lead in paint is recommended. (See "Testing Your Home for Lead" on the reverse side of this sheet for a brief description of the two basic testing methods most commonly used by professional inspectors.)

To receive a list of laboratories that have passed proficiency testing for analysis of lead in paint, soil, and/or dust samples, call the National Lead Information Center Clearinghouse at 1-800-424-LEAD (in the Washington, D.C., metropolitan area, call 202-833-1071). You may also fax your request for this list to 202-659-1192. When you call or fax, please specify that you are requesting the NLLAP/ELPAT List of Laboratories.

The federal government does not maintain a comparable list of X-ray fluorescence (XRF) operators. Some states may maintain their own lists, however.

6/93

The National Lead Information Center is operated by the National Safety Council with funding from the Environmental Protection Agency, the Centers for Disease Control and Prevention, the Department of Housing and Urban Development, and the Department of Defense.

FIGURE 5-2. National Lead Information Center fact sheet concerning home test kits for lead.

presence of lead-based paint, what matters most is that the amount of lead in the paint is significantly greater than the regulatory limit, which in the case of federally funded public housing authorities is 1.0 mg/cm^2. Currently available XRF lead paint analyzers are undergoing evaluation by various agencies and organizations for their ability to accurately measure lead concentrations at or near regulatory threshold levels; inspectors are urged to stay current of trade and scientific publications, which will report on the outcomes of these evaluations.

Survey testing is characterized by production of data that is of known accuracy and precision (less than from analytical testing), resulting from conduct of the instrument manufacturer's procedure and protocols. Results are element specific (again less than analytical methods) but must be corrected in some way for interferences. For XRF, methods include a procedure for verification of the as-manufactured instrument calibration, since instruments are not calibration adjustable.

Because they are conducted in the field, XRF tests are useful in that data are generated for right-now use. The XRF is used to "survey" the site (most commonly painted surfaces) for lead content. The data generated is used to determine where analytical testing is needed, or where the lead content is such that no further testing is required.

Screening Tests

Screening tests are very sensitive to the presence of lead, are used to respond to a yes/no question about lead content, and are highly precise around a transition or end point. They allow for verification of only one side of the yes/no response, usually considered to be the "yes" response.

Nearly all screening tests for lead are colorimetric tests, called spot tests. These spot tests most commonly use sulfide or rhodizonate to screen for the presence of lead. Sulfide and lead result in a black precipitate (solid) of lead sulfide, while rhodizonate and lead result in a pink/red precipitate of lead rhodizonate.

As these tests show color change at very low concentrations of lead, they can be used to test (screen) for the presence of lead. They can be used to answer the question "Is there lead present? Yes/No." Also, within some limits, adjustments to the chemistry of spot tests can be made to change the question to "Is there lead present at or above a target/regulatory concentration? Yes/No."

The use of chemical spot testing is either destructive or nondestructive. Some tests involve cutting the tested surfaces (a V incision) with a utility knife and applying the chemical to exposed underlying layers of paint. Other tests require removing a portion of the paint and performing the spot test away from the painted surface. Still others test only the intact surface of the paint.

Spot tests can be performed quite quickly, as the color change occurs within a few seconds or minutes, but interferences that may give errors need be known and considered. For example, sodium sulfide will give positives when used on painted steel surfaces (iron sulfide is black). The inspector should question the manufacturer and/or supplier about interferences and other limitations.

Chemical spot tests are very sensitive to the presence of lead and are rapid in comparison to the other methods available, but do not quantify the amount of lead present. Spot testing can be useful in the screening search for the presence of lead. Once found, further or other tests may be required before decisions may be made, or conclusions reached.

Spot tests are the least expensive to conduct but offer the least information as they respond to a yes/no question only.

ACCREDITATION AND CERTIFICATION

Of all the sampling test types (matrix specific) available to lead abaters, only some of those analyses performed in a laboratory have accreditation and/or certification programs to assure quality of conduct. The accreditation programs are listed in Table 5-1 as a function of the sample matrix. The federal government authority having jurisdiction is also listed.

Of survey and screening tests available, survey by XRF and screening using sulfide spot tests are predominantly recommended by governmental authorities.

Survey by XRF is recommended by the Department of Housing and Urban Development (HUD) for examination of residential property (dry paint film only). That agency recommends laboratory testing for verification of XRF results (dry paint film) or a method for lead concentration determination, and requires laboratory (analytical) testing for lead in other matrices (soil and settled dust).

Use of the sulfide spot test for screening dry paint film is recommended by some states but by no federal agency.

TESTING USE

Abatement requires testing before, during, and after the job. Abatement also requires monitoring (a testing-over-time process). Monitoring begins before the job starts and extends until the job is complete.

Preabatement Testing

Mostly this means that you wish to know if there is lead present, where it is, and how much (concentration) is there.

For the abatement of lead-based paint, most thought about testing focuses on testing paint film for the presence and concentration of lead. The center of attention should be to find the paint to be abated, but some thought needs to be spent on other matrices.

TABLE 5-1. Tested Sample Matrix versus Laboratory Accreditor and Federal Authority Having Jurisdiction

Matrix	Accreditor (Authority)
air, ambient	(EPA)
air, workplace	AIHA, A2LA (OSHA)
blood, public	CDC (HHS)
blood, worker	CDC (OSHA)
dust, settled (wipe)	AIHA, A2LA (EPA)
paint, dry film	AIHA, A2LA (EPA)
soil	AIHA, A2LA (EPA)
solvents	(EPA)
waste, by TCLP	(EPA)
water, drinking	state or EPA (EPA)
water, sanitary	local or state (EPA)
water, waste	(EPA)

Note: The accreditor is the organization that grants certifications or accreditations to a laboratory. The authority, as shown in parentheses, is the government agency that regulates the accreditation process and in some cases is the same entity as the accreditor.

EPA = Environmental Protection

AIHA = American Industrial Hygiene Association

A2LA = American Association for Laboratory Accreditation

OSHA = Occupational Safety and Health Administration

CDC = Centers for Disease Control and Prevention

HHS = Department of Health and Human Services

TCLP = Toxic Characteristic Leaching Procedure

A proper and conservative approach is to assure that liability is controlled. This is done by testing not only the paint to discover if it is lead-based but also the soil near and around the abatement site, the water near and plumbed into the abatement site, and the blood of the workers who will perform the abatement.

Testing the soil near and around the abatement site will record the concentration of lead already in the soil, for which the contractor is not responsible. If the soil at the site already has a high concentration of lead, the contractor surely wants to know. Obviously, after the abatement job is complete, the contractor will test the soil again to prove that he or she released no additional lead into it.

Testing the water (lakes, ponds, streams) near the site documents existing lead concentration for which the contractor is also not responsible. After the job, testing should show no increase as a result of the abatement.

Testing the water plumbed into the site checks for another source of lead. Some abatement contracts require abatement of all lead sources. This could be a source. Further, workers will be drinking this water. The contractor needs to know all lead sources to which the workers may be exposed.

Testing of workers prior to the job is used to document any change in blood lead level as a result of conduct of the job. The contractor/employer is responsible for the protection and health of workers on the job (OSHA). Testing the blood of workers before the job begins establishes a base-line against which changes can be observed. In addition, data on blood levels of occupants may reveal evidence of lead exposures for which the contractor is not responsible.

Monitoring During the Job

Occupational Safety and Health Administration regulations require that workers may not be exposed to lead dust in air above a certain limit called the permissible exposure limit (PEL). They further require that certain actions must be taken if the lead-dust exposure exceeds an action level. The only way to determine the exposure level is to test the dust in the air breathed by workers. This monitoring program is also called an industrial hygiene program or a worker monitoring program. Personal samples (collected on workers) and area samples (collected in the work space) are taken within (and possibly around) the workplace. Area dust monitoring is used to define the boundaries and extent of the workplace, also required by OSHA regulations.

The Environmental Protection Agency (EPA) requirements allow ambient air to contain limited amounts of leaded dust and limited amounts of dust in general. Monitoring may be required to assure that these levels are not exceeded.

Testing During the Job

As the job goes on, testing is done to check progress. The program could include continuing tests of paint film available for abatement, tests of dust generated as a result of abatement activities (dust in air as well as settled dust), and certainly tests must be done on waste created to determine if it is hazardous. Doing these tests while the job is going on may save much. For example, it could prevent contamination of nonhazardous waste with hazardous waste, thus reducing disposal cost by reducing the total amount of hazardous waste.

Postabatement Testing

As mentioned above, testing of soil and water near and around the site need be done to show no contribution from conduct of the job. Testing of worker blood lead levels completes the across-the-job testing used to protect the health and well being of the workers. When available, data on blood levels of occupants may be examined to determine levels increased during the course of the job.

Wipe tests for settled dust in residential abatement projects are used to determine clearance for reoccupancy. There are regulatory levels (EPA/HUD) to which public housing residences must comply prior to reoccupancy. Dust testing in other situations is done to show conduct of cleanup. Where levels are not established by regulation, dwelling reoccupancy values may be used as a guide.

X-RAY FLUORESCENCE ANALYZER (XRF) TEST PROTOCOLS

A Consumer Product Safety Commission (CPSC) staff analysis[3] of field grade XRF lead paint analyzers contains the following conclusions:

> The use of field grade XRF meters for the detection of lead paint in consumer homes should no longer be recommended. The field meters are not sufficiently reliable to measure lead-in-paint levels that can produce a substantial health hazard. . . .
>
> The HUD action level of 1.0 mg/cm^2 may be detectable by some XRF meters, but it is too high to adequately protect consumer health. Furthermore, a below-HUD action level reading by XRF could falsely indicate that special procedures and protective equipment would not be needed for the removal of some lead paints. This could result in the creation of lead dust hazards by consumers as well as professionals.

Another report that was released by a CPSC representative (but not reviewed by the Commission) states: "Field-grade XRF units typically do not give reliable readings below 2.0 mg/cm^2. Some leaded paints will not be detected."[4]

While these reports contain very general comments on the limitations of unspecified XRF analyzers and do not provide data to substantiate these limitations, they do raise concerns for firms conducting lead paint testing with XRFs.

The accuracy and precision of field grade XRFs is the subject of ongoing intensive evaluations by federal agencies. In the meantime, XRF users may wish to implement some interim measures to "qualify" their data until new guidance and testing specifications are issued by federal agencies.

One measure to consider is to increase the number of confirmatory paint chip samples from surfaces that test below 2 or 3 mg/cm^2, and to have the laboratory provide the analytical results in units of mg/cm^2. Thus the laboratory results will remain consistent with XRF values. To accomplish this requires careful measurement of paint chip size and good sampling technique to assure that the paint film has been removed down to the substrate.

It is important to consider that paint films with lead values below the HUD limits of 1.0 mg/cm^2 or 0.5 percent lead by weight are potentially hazardous.

Therefore it would be prudent to recommend that precautions to minimize, contain, and clean up dust be taken when paint films with any detectable amounts of lead are to be disturbed by renovation activities.

Where XRF testing methods and specific lead action limits (such as HUD's 1.0 mg/cm^2) are specified by housing agencies or other clients, the testing firm should consider appending to its report of results a statement that the firm makes no representation or warranty regarding the validity of action limits as protective of potential exposures to lead.

The remainder of this section provides an overview of HUD-mandated XRF testing methods applied in different types of housing. Lead-based paint testing is a young industry, and readers are advised to keep abreast of new methods and improvements to testing protocols that are forthcoming from federal agencies and professional organizations.

Section 302 of the Lead-Based Paint Poisoning Prevention Act of 1971 (LBPPA) requires that public housing authorities (PHAs) and Indian Housing Authorities conduct lead-based paint testing in dwelling units and common areas in all developments where children under the age of 7 live or are expected to live. However, individual state and local governments may set different requirements for lead-based paint testing. Therefore, inspectors must be aware of all regulations which may apply to their project. Inspections are generally classified into three categories: random, comprehensive, and determinative. Each type of inspection has a different methodology and purpose.

Random Inspections

Random inspection is used in multifamily housing developments. This type of housing development generally consists of apartment buildings forming a single complex. Typically the dwelling units in such developments are of similar construction and share a common maintenance program. Based on the history and maintenance records of the housing development, random inspection and sampling of the units is appropriate. Random inspection and sampling provides representative data for decision making without the cost associated with exhaustive testing. This method may also be applied to smaller housing developments, but it should be noted that as the size of the development decreases the percentage of units to be tested increases when using random selection methods.

Inspection Plan

Random inspection in a multifamily development requires the formulation of a Lead-Based Paint Inspection Plan that may contain a listing of the specific dwelling units, common areas, and exterior areas to be inspected and a delineation of the specific architectural components to be tested in each unit, each common area, and on each building exterior. In order to develop such a plan, the following is required:

1. A complete listing of all units and common areas in the development.
2. For each unit type within the complex, a list or drawing showing all painted surfaces, both interior and exterior. These documents must show layouts of all the rooms, and show what architectural components are contained in each of the rooms (walls, window sills, doors, moldings, etc.)
3. Listings or drawings of common areas and components.
4. Information on building types and typical exterior components.

If a housing manager has knowledge of lead-based paint, based on construction history, or previous modernization work, the inspector should be informed. Information provided by the housing manager on the construction history and known presence or absence of lead-based paint in a project can be used to tailor the sampling strategy. Building materials installed after 1978 need not be tested for lead since lead-based paint was banned for residential use in that year, HUD has stated. If a housing manager has knowledge of post-1978 modernization/replacement of components, then those components need not be tested.

The HUD Interim Guidelines suggest that a subsample of units within a housing development be inspected rather than every unit in the development. This concept is based on the idea that developments are uniformly painted at the time of construction, and that subsequent painting of units followed a common protocol. The key to effective random inspection is grouping units with similar construction dates and painting histories. The HUD Interim Guidelines call this grouping of units "groups of similar buildings."

Once a group of similar buildings has been established, the number of units in that group must be determined. Table 5-2 is reproduced from the HUD Interim Guidelines. The first column indicates the number of units in a building or group of similar buildings. The second column indicates the number of units to be sampled based on the total number of similar buildings.

The next step in developing the inspection plan is determining which specific units are to be inspected. If a building contains 80 units, 42 units are to be inspected. These units must be selected randomly. A simple way to accomplish this is to select units from the list at a fixed interval. For example, 42 units must be tested in an 80 unit development, based on Table 5-2. Since 80 divided by 42 equals 1.9, every other unit (second) on the list should be tested. This method of random generating may produce an obvious bias sample of units (all even or odd numbered, all vacant, or all two-bedroom units). If this bias occurs, the problem may be solved by starting the sample selection at a different unit.

Within each dwelling unit sampled, a representative sample of each type of painted component in every room, closet, pantry, hall, and so on must be tested. Table 5-3 below lists the HUD Interim Guidelines surface testing sites. This list should not be considered complete, but rather a guide to components generally found. If unlisted components are encountered they must be tested. Note that this

TABLE 5-2. Number of Units To Be Tested as a Function of Project Size

Number of Units in Building or Group of Similar Buildings	Number of Units to Be Tested
20	All units
40	31
60	38
80	42
100	45
200	51
300	54
400	55
600	56
1000	57
1500	86
2000	115
2500	144
3000	174
3500	203
4000	232
4500	261
5000	290
>5000	299

Source: HUD Interim Guidelines, 1991.

Note: For intermediate project sizes between 1000 and 5000, the number of units to be sampled may be approximated by adding 6 sampled units per 100 units in the project to the previous sample size in the table. For example, the number of units to be sampled in a 2700 unit project is $144 + 2 \times 6 = 156$.

table refers to components with the same painting history. Thus if a room has two doors of different color, both doors must be tested. Refer to Figures 5-3 through 5-8 for illustrations of surface testing sites.

Substrate Interference

Once a sampling plan has been established and prior to the commencment of XRF testing, substrate interference must be accounted for. Typically, the first two or three units scheduled for testing are used as protocol units. At this time, substrate corrections are determined. The term *substrate equivalent lead* (SEL) is used to describe the "substrate correction" for the direct reading analyzer. HUD recommends substrate correction using spectrum analyzers only when the inspector deems it necessary and appropriate to do so. A procedure to determine whether substrate correction is properly functioning is described by the equipment manufacturer.

When using direct read XRF analyzers, the displayed lead reading taken on a painted surface is termed an *apparent lead concentration* (ALC), due to the possibility that the substrate and other factors may contribute to the ALC in a variety of

TABLE 5-3. Surface Testing Sites

Interior: In *each area* (each room, closet, pantry, hall, part of a divided room such as the dining area of a kitchen/dining room, etc.), the following surfaces are suggested for screening:

Baseboard	1 in each area
Ceiling	in each area
Crown molding	1 in each area
Door	surface of door and one side of the frame on a representative interior door in each area
Fireplace	—
Floor	1 in each area
Radiator	—
Shelf	1 in each area
Shelf support	in each area
Stairs	riser, tread, stringer, newel post, railing cap, balustrade, upper wall, lower wall, and chair rail (if applicable) in each area
Wall	upper wall, lower wall, and chair rail (if applicable) in each area
Window	sash, casing, and sill on a representative window
Exterior	
Bulkhead	—
Ceiling	—
Cornerboard	—
Door	surface of door and door casing
Fence	—
Floor	—
Joist	—
Lattice	—
Lower railing	—
Painted roofs	—
Porch	—
Railing cap	—
Siding	—
Stairs	tread, riser, and handrail
Support column	—
Trim	upper and lower
Window	sill, casing, and sash of a representative (also cellar window unit)

Source: HUD Interim Guidelines, 1991.

ways. The inspector must compensate for this possibility by measuring the contribution of the substrate to the apparent lead readings. This substrate contribution is termed the SEL of the specific building component substrate. The average SEL is measured by removing the paint from the substrate and taking XRF readings on the substrate where the ALC was obtained. The resulting corrected lead concentration (CLC) is the difference between the average ALC and the average SEL (CLC equals ALC minus SEL).

Once the SEL has been determined for a component's substrate, it will be subtracted from all like components tested throughout similar buildings.

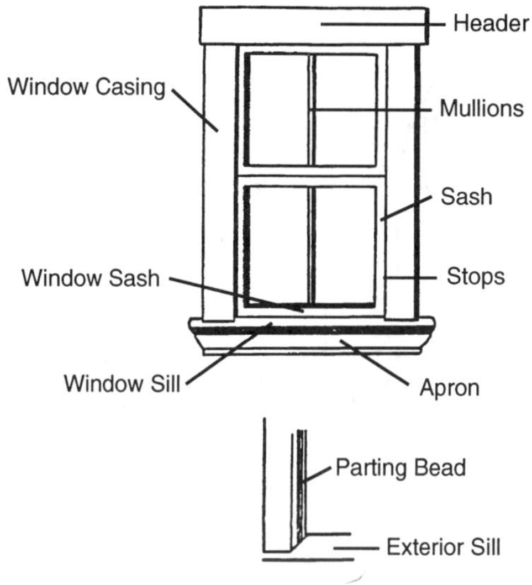

FIGURE 5-3. Window area surfaces for lead testing.

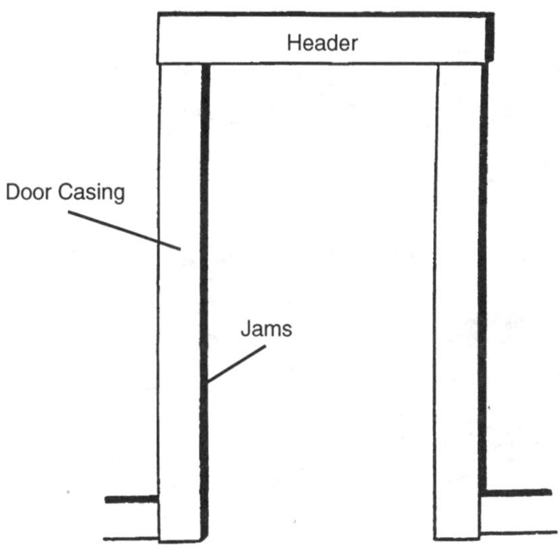

FIGURE 5-4. Doorway surfaces for lead testing.

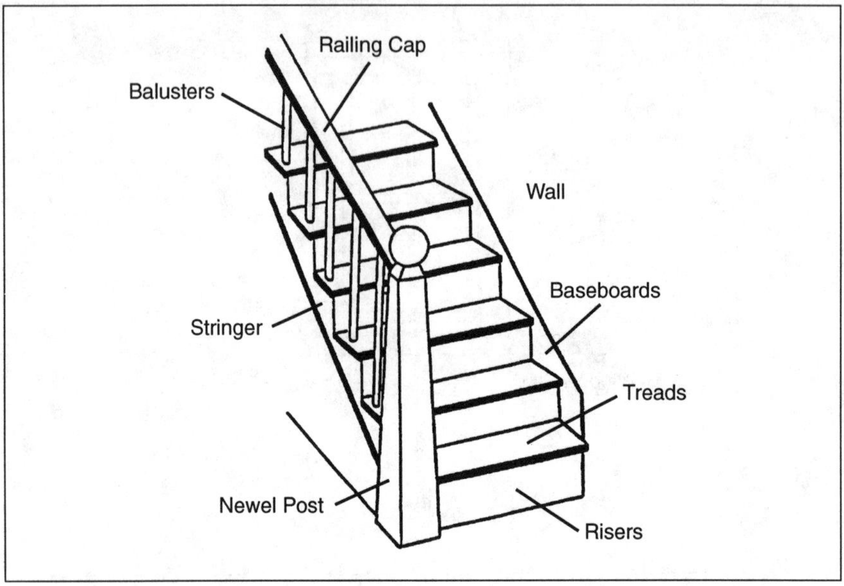

FIGURE 5-5. Stairway area surfaces for lead testing.

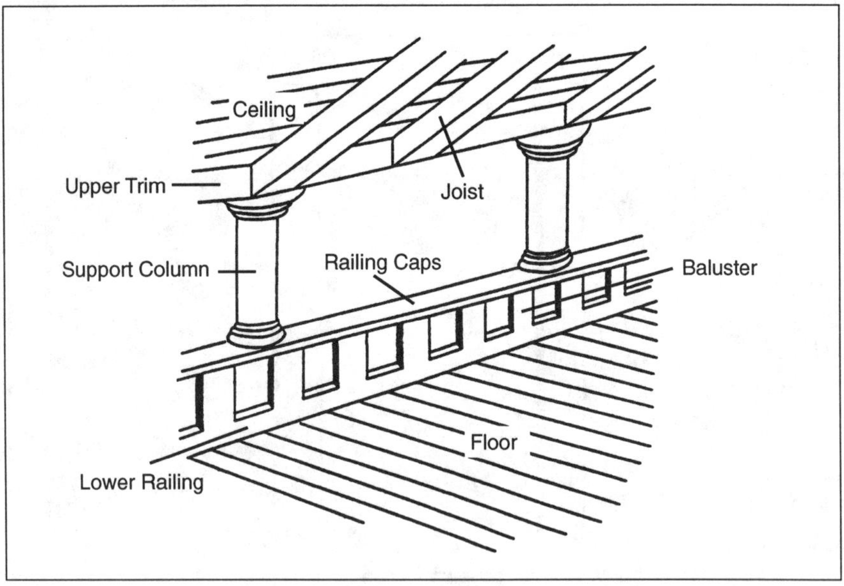

FIGURE 5-6. Porch surfaces for lead testing.

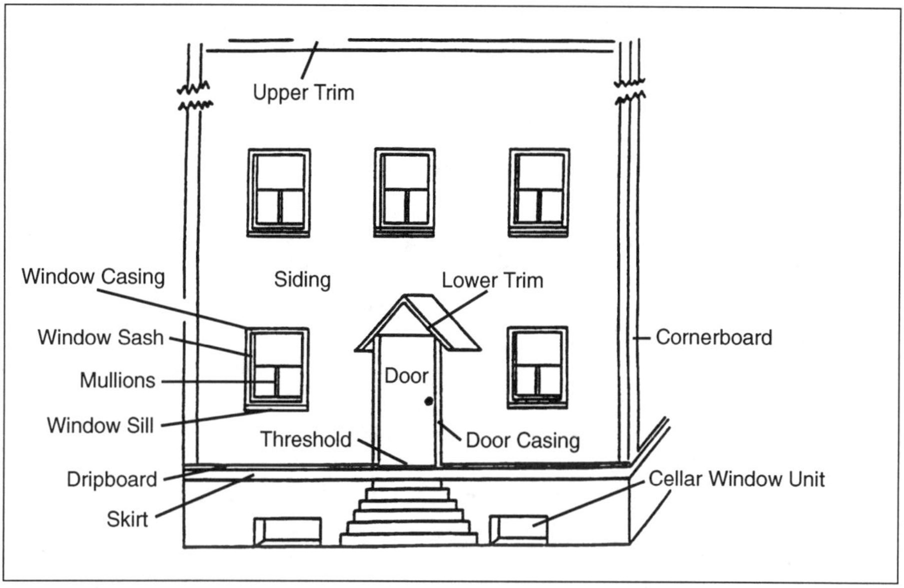

FIGURE 5-7. Exterior building surfaces for lead testing.

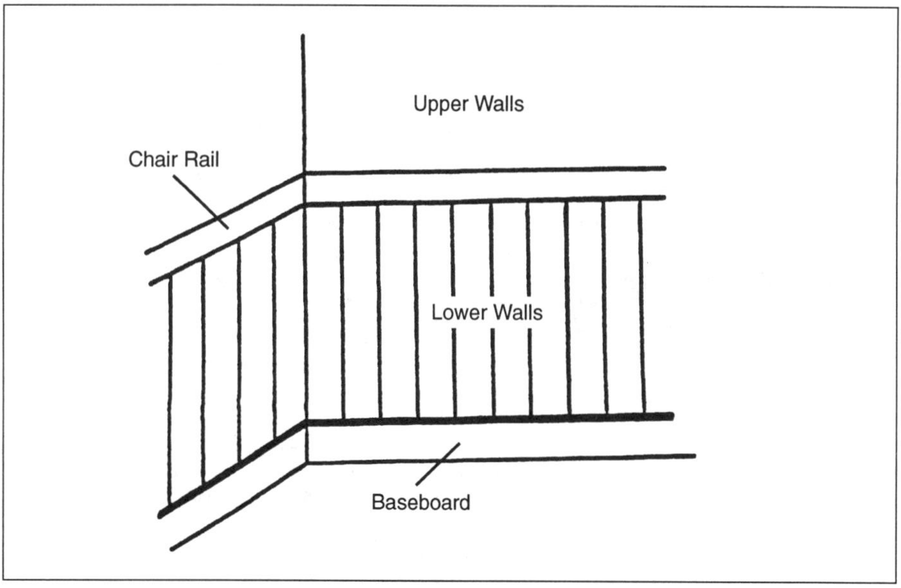

FIGURE 5-8. Interior wall surfaces for lead testing.

Classification of Components

According to the HUD Interim Guidelines, components are determined to be positive if the amount of lead in the paint is equal to or greater than 1.0 mg/cm^2. Because there is some uncertainty in measurements obtained from portable XRF analyzers, the criterion for classifying components is based upon historical data about error size and on the principles found in the science of statistics.

The outcome of XRF testing is the finding that a component within a building or project will be classified as *positive, negative,* or *inconclusive* for lead-based paint. Classification rules to reach a finding differ, dependent on the type of portable XRF analyzer in use by the inspector. The rules for classifying components using the direct reading XRF instrument are outlined in Table 5-4. The rules for classifying components using the spectrum analyzer are outlined in Table 5-5.

When using either the direct reading or spectrum analyzer, the term *inconclusive* is used to suggest that further testing, possibly using another analysis technique, may be needed to determine if lead-based paint is present. Sample results within the range considered inconclusive usually require confirmatory laboratory analysis.

Each measurement should be recorded to indicate whether that component is classified as positive, negative, or inconclusive. At the end of the testing process these results will be tabulated to determine abatement strategies. The summary of XRF testing data are usually presented to housing authorities in the format shown in Figure 5-9.

Interpretation of XRF Test Data

Once XRF testing data is compiled, conclusions about abatement are made using decision-making rules set out in the HUD Guidelines. The rules are applied to

TABLE 5-4. Rules for Direct Reading XRF Instruments: Multifamily Unit Housing

1. A CLC of 1.6 mg/cm^2 or greater indicates that the components is positive.
2. A component on which the CLC is below 0.5 mg/cm^2 is to be classified as negative.
3. A component on which the CLC is greater than or equal to 0.5 mg/cm^2 and less than or equal to 1.5 mg/cm^2 is classified an inconclusive.

Source: HUD Interim Guidelines, 1991.

TABLE 5-5. Rules for Spectrum Analyzer XRF Instruments: Multifamily Unit Housing

1. A value of 1.3 mg/cm^2 or greater (corrected if needed) indicates that the component is positive.
2. A component on which the value is below 0.8 mg/cm^2 (corrected if needed) is to be classified as negative.
3. A component on which the value is greater than or equal to 0.8 mg/cm^2 and less than or equal to 1.2 mg/cm^2 (values corrected if needed) is classified as inconclusive.

Source: HUD Interim Guidelines, 1991.

Building Component	Number Tested	Number Positive	Number Negative	Number Inconclusive
Baseboard	55	10	15	30
Walls	55	0	47	8
Steel Windows	55	3	35	17
Wood Windows	—	—	—	—
Wood Doors	—	—	—	—
Steel Doors	—	—	—	—

FIGURE 5-9. Summary of XRF testing data.

each component separately, depend on the type of XRF analyzer used, and are made dependent upon the percentages of the component classified as positive, negative, or inconclusive. The rules are to be applied in sequence until a decision is reached, and are as follows:

Rule 1: If more than the percentage of XRF results given below for a component are *positive,* lead is considered to be present, and either all such components should be abated, or all should be tested to determine which require abatement and which do not. If the testing option is chosen, all inconclusive results must be confirmed.

Percentages: 15 percent of direct reading XRF results or 11 percent of spectrum analyzer XRF results.

Rule 2: If no XRF results for a component are positive, and less than or equal to the percentage given below are inconclusive, no abatement or further testing of that component is required.

Percentages: 17 percent of direct reading XRF results or 0 percent of spectrum analyzer XRF results.

Rule 3: If some XRF results, in the percentages given below, are positive or inconclusive, confirmatory testing is required.

Percentages: Direct reading XRF results are less than or equal to 15 percent positive or more than 17 percent inconclusive. Spectrum analyzer results are less than or equal to 11 percent positive or more than 0 percent inconclusive.

All positive XRF results must be confirmed. All inconclusive readings above 1.0 mg/cm^2 must also be confirmed.

Rule 4: If confirmatory testing under Rule 3 confirms the presence of lead-based paint above the regulatory level on a component, the public housing authority manager has the same options as under Rule 1: All components of the given type can be abated, or exhaustive testing of that component can be carried out to determine precisely which ones must be abated.

The following rules apply for abatement decisions in single-family housing when a direct reading XRF analyzer is used along with substrate correction:

1. ALC results of greater than or equal to 3.0 mg/cm^2 taken on flat surfaces require abatement.
2. ALC results of greater than or equal to 4.0 mg/cm^2 taken on nonflat, or curved surfaces or surfaces that are smaller than the sensitive area of the XRF used require abatement.
3. CLC values greater than or equal to 1.6 mg/cm^2 require abatement with no further confirmation.
4. CLC values less than or equal to 1.5 mg/cm^2 require confirmatory testing.

The following rules apply for abatement decisions in single-family housing when a spectrum analyzer XRF is used. All spectrum analyzer results should be reported with the substrate correction (if needed). In general, less confirmatory (laboratory) testing will be required when using the spectrum analyzer.

1. Spectrum analyzer values greater than or equal to 1.3 mg/cm^2 require abatement.
2. Spectrum analyzer values less than or equal to 0.3 mg/cm^2 indicate no lead is present at or above the regulatory level of 1.0 mg/cm^2.
3. Spectrum analyzer values between 0.4 and 1.2 mg/cm^2, inclusive, require confirmatory testing.

Determinative Inspection

This type of inspection involves determining the presence of lead-based paint on a very limited number of surfaces. This type of inspection should not be used in the place of random or comprehensive inspection. Lead abatement should not be performed on a residential property in which only a determinative inspection has been performed.

The state of Massachusetts recognizes two types of lead-based paint surveys: (1) a lead determination and (2) a comprehensive inspection. The "lead determination" is a survey used only to determine whether or not lead-based paint is present in the residence. This survey may not be used to decide whether to conduct abatement operations, nor may it be used to achieve "compliance." The "comprehensive inspection," on the other hand, requires that all accessible and mouthable surfaces 5 feet (from the floor) or lower be tested for the presence of lead-based paint. In addition to these surfaces, all loose, peeling, and flaking paint, regardless of height above the floor, must be tested. Contact or friction surfaces, such as a door frame or in a window channel, must also be tested regardless of height. This comprehensive survey may be used to decide to perform lead abatement work.

Determinative inspections may also be used outside of the residential setting. Often selected surfaces require lead-based paint testing prior to construction or demolition of commercial properties. If lead is found on component, OSHA and EPA regulations will need to be referenced, concerning worker protection and classification of hazardous waste disposal.

Comprehensive Inspection

Inspection of single-family or scattered site housing differs from inspection in multifamily units. The decision to abate is based on sampling in that single unit. Since only one unit is involved, comprehensive testing must be performed. In these situations, the HUD guidelines suggest that all surfaces be tested.

The surfaces to be tested in single-family dwellings and in scattered site housing are the same, but not limited to, those suggested for testing in multifamily units. It is important to keep in mind that the HUD guidelines only pertain to public and Indian housing units. Nonetheless, it is suggested that owners of single family units follow them to ensure that testing and inspection and testing of their dwelling is conducted properly.

Notes

1. Draft proposed standard, Occupational Exposure to Lead in Construction Operations, Department of Health Services/Department of Industrial Relations, State of California, May 11, 1992. Cited in *Federal Register,* May 4, 1993. Lead in Construction, Interim Final Rule, 29 *CFR* Part 1926, May 4, 1993, p. 26590–26649.
2. Ibid.
3. CPSC Staff Analysis. *Screening Lead Paint in Homes by Field Grade X-Ray Fluorescence Meters,* May 14, 1993.
4. Consumer Product Safety Commission, Directorate for Health Sciences, Lead Poisoning Project (Brian Lee, manager). *Screening Lead Paint in Homes by Field Grade X-ray Fluorescence Meters,* October 27, 1993.

6

Risk Assessment, Part 1: Purpose and Definitions

David Jacobs, C.I.H.

Risk assessment is a process that identifies lead hazards of immediate concern and provides advice on how to control them. Originally developed for insurance purposes in public housing, lead-based paint risk assessments will be conducted with increasing frequency in the nation's private housing. Conducted properly, risk assessment will not only point out what kind of lead hazards exist in a property but will also reorient the owner's view of deteriorating paint and dust. Paint deterioration was previously considered to be a "nuisance" housing problem. However, risk assessors are in an ideal position to provide education and insight on why mere cosmetic approaches to addressing the problem will be inadequate.

HAZARD IDENTIFICATION

The first step in developing a lead hazard control plan for a property is to identify the presence or absence of lead hazards. Bypassing the hazard identification process can result in spending money correcting hazards that do not really exist, or even worse, failing to correct problems that go undetected, resulting in a lead poisoned child. There are two forms of hazard identification in the lead-based paint field: risk assessments and inspections.

Although the two terms are sometimes used interchangeably, inspections and risk assessments are defined quite differently under statute (see Title X of the Housing and Community Development Act of 1992 [P.L. 102-550]). When contacted by an owner, the risk assessor's first step should be to help the owner determine whether an inspection and/or risk assessment is appropriate.

Risk Assessment

A risk assessment provides information and advice to the owner of a dwelling unit so that quick action can be taken to control lead hazards posing immediate danger to occupants or maintenance workers. In order to provide the necessary guidance, a risk assessment is composed of three parts:

1. Identification of the existence, nature, severity, source, and location of immediate lead-based paint hazards
2. Identification of some of the owner's management and maintenance practices that may require modification to address lead-based paint
3. A determination of the various options for feasible hazard control, including interim control and abatement measures, if hazards are identified

Since there may be some lead-based paint left in the dwelling after some types of interim controls and abatement, the risk assessor should also teach the owner how to keep that paint in a nonhazardous condition over the lifetime of the building.

The risk assessment protocol contained in this chapter is based partly on earlier work in public[1,2] and private housing. The protocol described in this chapter represents the minimum standard for conducting risk assessments. This protocol has attempted to strike a balance between the need to provide information of sufficient reliability and the need to keep costs low. More elaborate and extensive risk assessments may be needed in certain situations. For example, the protocol described here does not replace or supplant the public housing protocol, which is designed to meet the more complex management and maintenance needs of public housing authorities. An investigation of a residence housing a child with an elevated blood lead level will also require a more elaborate protocol in accordance with local health agency requirements. This protocol is focused on the condition of the housing, not the health status of a child. Fundamentally, risk assessments constitute a proactive primary prevention strategy.

Lead-Based Paint Inspections

Lead-based paint inspections can be performed by either a certified technician or a certified risk assessment professional. Some jurisdictions require a license to perform these functions. Inspections measure the concentration of lead in paint on a surface-by-surface basis. Inspector technicians can also collect surface dust and/or soil samples for clearance purposes following interim controls or abatement. However, an inspection does not necessarily develop a set of recommendations on what to do, or teach the owner how to keep lead-based paint in a nonhazardous condition, since management and maintenance data are not considered during inspections.

TABLE 6-1. Comparison of Risk Assessment and Inspection

Analysis	Risk Assessment	Inspections
Defective paint *only*	Required	No
Dust	Required	Optional
Soil	Required[a]	Optional
Maintenance	Required	Not required
Management	Required	Not required
Status of any current child lead poisoning cases	Required	Not required
Review of previous paint testing	Required	Not required
All surfaces with different painting histories	Not required	Required
Typical applications	1. Vacancy 2. Bldg. nearing the end of expected life 3. Limited funding available 4. Sale of property 5. Insurance	1. Abatement 2. Rehab work 3. Weatherization 4. Sale of property
Final report	Lead hazard control plan	Lead concentrations for each surface tested

[a]Unless previous reliable neighborhood sampling has already established that lead in soil levels are well below 500 ppm.

Table 6-1 summarizes some of the key differences between risk assessments and inspections.

Combining Risk Assessments and Inspections

It is possible—indeed, preferable—for an inspection and a risk assessment to be done simultaneously. By combining measurements of dust and soil with surface-by-surface paint analysis and by collecting maintenance and management data, both immediate and potential lead-based paint hazards can be identified and addressed in a comprehensive fashion for the life of the building, employing the best mix of interim control and abatement strategies for any given owner(s).

If an inspection has already been completed before a risk assessment has occurred, the risk assessor will evaluate the inspection report to determine if the information is sufficiently reliable to guide proper management of lead-based paint. If an inspection has not yet occurred, a risk assessor will typically recommend that all untested old paint applied before 1978 be presumed to contain lead above the regulatory limit until an inspection can be completed. In short, inspections and risk assessments are complementary but independent activities.

It is up to the owner to decide which type of hazard recognition effort is appropriate, unless certain federal subsidies are being used; in some cases, subsidies will require that a risk assessment or an inspection be completed. Since risk assessments characterize immediate hazards, they should be conducted when the owner wants to know only what should be done in the short run. For example, buildings that are expected to be demolished in several years might benefit most from a risk assessment, since abatement will probably not be performed. On the other hand, inspections will enable the owner to manage *all* lead-based paint by determining exactly where the lead-based paint is. However, an inspection could miss immediate hazards that could be recognized by a risk assessment.

THE RISK ASSESSMENT PROCESS

Hiring a Risk Assessor

The risk assessor is hired by the owner or is employed by the public sector, but in any case is an independent, trained, and licensed (in states where licensure requirements have been promulgated) professional who provides an objective analysis of the lead hazards found in a property. Some landlords could have a management staff person trained and licensed to aid in the decision-making process, but such an assessor may not be perceived as being able to provide a truly independent evaluation of the property. Therefore, the owner should contract with an independent, professional risk assessor to minimize the perception of bias, which would be especially important in the event of litigation. Risk assessments can be conducted when a dwelling becomes vacant, since abatement or interim control measures can be implemented most easily at that time. However, risk assessments can also be completed in occupied units, depending on how owners wish to phase in hazard identification programs in their dwellings.

Exchanging Information

Typically, the owner will provide to the risk assessor information on previous lead-based paint abatement activities (if any), management and maintenance practices, any existing cases of lead poisoning at the site (if known), age of dwelling, occupant information, and other information. This is evaluated in all kinds of ownership settings, including individual homeowners, small landlords, property management companies, and large landlords.

Developing a Site Visit Plan

The risk assessor then reviews these data to focus the limited environmental sampling and visual assessments and to identify key individuals who will be involved. If the risk assessment is being performed in multifamily housing with 10 or more similar units, the specific units to be sampled will then be selected. If the population of dwellings are varied, then each will require a separate risk assessment, although it may be possible to make certain exceptions to this rule.

Visual Assessment and Environmental Sampling

A visual assessment of each sampled unit is conducted at this time. Defective paint, dust, and soil samples are then collected and submitted to a laboratory for analysis. If there is an adequately sized intact surface, portable X-ray fluorescence (XRF) can be used for the defective paint analysis. All layers of paint should be analyzed, not just the deteriorated layers at each testing location.

Interpreting Data and Preparing a Report

When the lab results are received, the risk assessor reviews all data, including visual assessment results, environmental sampling results, and management and maintenance information. A report listing various lead hazard control options, including a broad spectrum of treatments ranging from interim controls to full abatement for all identified lead hazards, is drafted. The report includes estimated costs of each option, both in the short run and the long run. The recommended treatments will be based on the risk assessor's professional judgment of the property owner's management and maintenance capabilities, the condition of the property, and the location and severity of lead-based paint hazards.

Implementing a Lead Hazard Control Plan

Typically, the owner selects a particular lead hazard control strategy, implements it, and, depending on local regulations, is provided with a certificate or report by either the risk assessor or local government after the strategy has been successful in complying with the applicable standards.

LIMITATIONS OF THIS RISK ASSESSMENT PROTOCOL

Assessment of Less Common Sources of Lead Exposure

In order to evaluate the largest number of dwellings in the shortest period of time, this risk assessment protocol does not examine *all* potential sources of lead in each dwelling. Instead, this protocol examines the *most likely* sources of lead hazards that can be corrected by a property owner. If it is known that there are sources of lead poisoning peculiar to a given geographical area, those sources may also be assessed as part of primary prevention efforts, although probably not by a privately employed risk assessor. For example, if it is known that the use of folk remedies containing lead is widespread in a given neighborhood, the local health department will take appropriate steps. This protocol does not require private risk assessors and building owners to undertake public sector responsibilities that lie beyond controlling lead exposures not directly related to the residence itself.

Similarly, air sampling will be uninformative and is therefore not recommended for routine risk assessments for housing. If there is an identifiable lead air emission source nearby, public sector agencies will be responsible for correcting it, not private risk assessors.

Water sampling is also not recommended on a routine basis under this protocol, since the Environmental Protection Agency (EPA) already has an on-going program in this area that does not require duplication. However, risk assessors can collect water samples for an additional fee if requested by an individual homeowner or landlord.

Risk Assessment and Investigation of EBL Children

The risk assessment protocol contained in this chapter is not appropriate for investigation of a dwelling presently housing a child with an elevated blood lead level (EBL). For this situation, investigation of all sources may be needed. For a child with an elevated blood lead level, it is quite possible that a relatively uncommon source (e.g., glazed pottery or leaded toys) or even another dwelling is the source of contamination that requires identification.

Comprehensive Exposure Models

Computer exposure or risk assessment models that integrate various exposure sources and pathways are not recommended for routine residential risk assessments for the following reasons:

- Sampling of all sources of lead in millions of dwellings is not feasible
- Routine use of a complex computer model in millions of dwellings is not feasible
- There is little agreement within the scientific community on which model best characterizes risk at the present time.

The Environmental Protection Agency has developed a computer model that seeks to integrate all lead sources in a given environment.[3]

BENEFITS OF RISK ASSESSMENT

By conducting a risk assessment and implementing a lead hazard control program, an owner will be able to demonstrate that reasonable preventive measures have been taken. The risk assessment will provide a rational basis for focusing control activities. Some jurisdictions or insurance companies may require risk assessors to provide certificates or other documentation after an owner has completed lead hazard control efforts. If no immediate lead hazards are identified, the owner may be awarded a certificate of lead safety, which may last two years. In any case, there will be a report documenting the findings. Some jurisdictions will issue a lead-free certificate only after completing an inspection of all painted surfaces, dust, and soil with levels below applicable standards.[4] If the lead hazards uncovered by a risk assessment are corrected, a similar certificate with an appropriate expiration date may be issued. This documentation will prove to be invaluable if a resident child with an elevated blood lead level appears. An up-to-date lead-safe certificate will provide evidence of an owner's good faith efforts to control hazards, although a repeated investigation may still be needed to determine if a new lead hazard has appeared since the last risk assessment.

Finally, and perhaps most importantly, risk assessments are a feasible way of moving away from sole reliance on blood lead screening programs to a system that focuses on correcting hazards before they actually cause a lead-poisoned child. The development of this primary prevention thrust is at the heart of the federal Residential Lead-Based Paint Hazard Reduction Act of 1992 (Title X).

COSTS OF RISK ASSESSMENT

Because this protocol is limited to identifying the most likely lead hazards that can be controlled by owners, it can be implemented at relatively low cost. Since risk assessments permit owners to focus their responses on true immediate hazards, the cost of a risk assessment can usually be recovered by not wasting money controlling lead hazards that do not really exist.

Since all units with similar characteristics need not be sampled, the cost of conducting risk assessments on a dwelling unit basis in multifamily housing can be substantially less.

Notes

1. Housing Environmental Services/Housing Authority Risk Retention Group. *Lead-Based Paint Risk Assessment Protocol.* Cambridge, MA, 1991.
2. Department of Housing and Urban Development, Office of Public Housing. Notice of fund availability for lead-based paint risk assessments. *Fed. Reg.* 57(125):28910–28943.
3. U.S. Environmental Protection Agency. *Review of the National Ambient Air Quality Standards for Lead: Exposure Analysis Methodology and Validation.* EPA-450/2-89-011. Research Triangle Park, NC, 1989.
4. State of Rhode Island and Providence Plantations Department of Health. *Rules and Regulations for Lead Poisoning Prevention,* R23–24.6–PB, 1993.

7

Risk Assessment, Part 2: Hazard Recognition Procedures

David Jacobs, C.I.H.

The on-site phase of risk assessment involves collection of a limited number of dust, soil, and defective paint samples. A visual assessment is also conducted at this time to determine the extent of visible dust accumulation and the existing condition of the paint. Standard field sampling forms for the on-site field testing are provided in this chapter. Revised Department of Housing and Urban Development (HUD) guidelines may provide slightly different forms that may need to be used in some jurisdictions.

VISUAL ASSESSMENT

The risk assessment will include an assessment of the condition of the painted surfaces, the quality of maintenance and housekeeping, and the presence of other substandard housing conditions. The existence of substandard housing conditions has been shown to correlate with elevated blood lead levels.[1] Sticky doors, windows, and other friction and impact points should be identified, since these may generate lead dust.

Condition of Painted Surfaces

The condition of painted surfaces can be evaluated by the risk assessor on Form 7-1. The risk assessor should rate each major type of painted building component as good, fair, or poor using the criteria in Table 7-1 (page 97). The risk assessor should compare this rating with that of the owner as a means of instructing the owner on current standards of care with regard to paint quality. If the paint on certain components is known not to contain lead above the regulatory limit, its condition need not be evaluated, although the assessor should recommend that it be repaired since many paints may have trace amounts of lead. Although most dwellings exhibit some minor degree of paint deterioration, it is commonplace for building owners to rate the condition of their paint more highly than a trained objective professional.

There are 12 paint conditions that can be considered "defective".[2] These conditions are:

1. Alligatoring: Reptilian scale patterns on dried paint films that are often caused by the inability of the topcoat to bond smoothly to a glossy coat underneath or by an excessively thick paint film.
2. Blistering: A formation of a bubble in the paint film caused either by heat or moisture. The risk assessor should break open one of the bubbles. If bare wood shows, then the likely cause is moisture. However, if another layer of paint shows instead of wood, heat probably caused the blister, not moisture. If moisture is suspected, the risk assessor should endeavor to locate the moisture source. Controlling the moisture source is a prerequisite for many forms of lead-based paint hazard control.
3. Checking: A pattern of short, narrow breaks in the top layer of paint, usually caused by a loss of elasticity in the paint. Plywood substrates can often cause checking.
4. Cracking and flaking: An advanced form of checking that usually occurs on surfaces with many layers of paint. The cracks usually form with the grain.
5. Chalking: A formation of a fine powder on the surface of a paint film, usually caused by a failure to adequately prime or seal a porous surface, overthinning of paint, or exposure to sunlight. Almost all exterior oil paints are designed to eventually chalk to wash dirt away in the rain and provide a good surface for repainting. The chalk may contain high levels of lead.
6. Scaling or cornflaking (peeling): A form of paint separation often forming in those areas of the building susceptible to condensation, such as under eaves. Deposition of salts drawn to the paint film surface by moisture or contact with sulfur dioxide can cause scaling.
7. Separation from asbestos shingles (peeling): A form of paint separation caused by chalk formation or painting over oil or dirt.
8. Separation from metal (peeling): A form of paint separation usually caused by improper priming of bare galvanized metal or by rusting, often seen on garage doors. The risk assessor should look on the backs of the paint chips. If rust is seen, the rust should be removed from the surface before recoating.
9. Peeling from hardboard siding: Caused by water vapor trapped in the paint film; moisture often comes from tightly constructed houses where moisture travels through the walls. The risk assessor should recommend that vents be installed or that worn caulk be replaced to eliminate the moisture problem.
10. Peeling from masonry and brick: Caused by salt crystal formation, which

pushes the paint away from the substrate. This condition is often caused by extraneous water migration at ground level due to poor drainage. The risk assessor should recommend proper drainage before attempting to coat this surface.

11. Peeling from wood: Usually caused when wet wood swells under paint, causing the paint film to loosen, crack, and fall off. The water can be caused by joints that are not caulked, worn-out caulking, ice-filled or plugged gutters, unvented buildings (which forces water vapor out through the walls), damp basements, leaking roofs, painting over wood while it is wet, and vegetation giving off moisture too close to the wood. The risk assessor should recommend installation of new caulk, vent installation, attic louvers, exhaust fans, dehumidifiers, gutter cleaning, removal of vegetation near painted wood, and so on, as a prerequisite to paint film stabilization and encapsulation.

In addition to determining the type and cause of paint deterioration, the risk assessor should estimate the extent of paint deterioration of each major building component type in the dwelling as indicated in Table 7-1. This information should be recorded on Form 7-1.

Condition of Building

Together with the history of code violations, the condition of the building is assessed as a means of determining the quality of the existing maintenance program, which will help the risk assessor determine the type of lead hazard reduction program that can be implemented realistically. The assessment of the building condition is also a way of determining the probability of future moisture damage to painted surfaces, which will determine the frequency of reevaluation required. A score for the building condition is assigned and can be used to adjust the repeated

TABLE 7-1. Categories of Paint Film Quality

Type of Building Component	Good	Fair	Poor
Components with large surface areas (walls, ceilings, floors, doors)	Less than ½ square foot	½ square foot to 2 square feet	More than 2 square feet
Components with small surface areas (window sills, baseboards, trim)	Less than 1% of the total surface area of the component	1–10% of the total surface area of the component	More than 10% of the total surface area of the component

Note: Building component in this table means each individual wall, ceiling, or floor, *not* the combined surface area of all similar components in a room.

FORM 7-1. Paint Conditions on Selected Surfaces (single family, owner-occupied)

Building Component	**Paint Condition: Good, Fair, Poor, or Not Present** (to be completed by risk assessor)	**Deterioration Due to Friction or Impact?**	**Deterioration Due to Moisture?**	**Location of Painted Component with Visible Bite Marks**
Building siding				
Exterior trim				
Exterior windows				
Exterior doors				
Railings				
Porch floors				
Other porch surfaces				
Interior doors				
Ceilings				
Walls				
Interior windows				
Interior floors				
Interior trim				
Stairways				
Radiator (or radiator cover)				
Kitchen cabinets				
Bathroom cabinets				
Other surfaces				

Note: If the overall condition of a component is similar throughout a dwelling, that condition should be recorded. If a component in a couple of locations is in poor condition, but the overall condition is good or fair, the specific sites of the badly deteriorated paint should be noted. The specific locations of any component with bite marks should be recorded.

monitoring schedule for specific levels of interim controls and hazard abatements, depending on local requirements. A building in poor condition will generally require more frequent reevaluation than will a building in better condition.

PAINT SAMPLING

Defective Paint

Defective paint constitutes an immediate lead-based paint hazard that should be evaluated as part of the risk assessment process. If an adequate paint inspection has not been completed previously, or is not being completed as part of a concurrent lead-based paint inspection, building components that exhibit chalking, peeling, flaking, or otherwise deteriorated paint should be analyzed for lead. One sample should be collected from each similar building component with similar painting histories. Paint analysis can be done either by laboratory paint chip analysis or portable X-ray fluorescence (XRF) analyzer. An XRF analyzer can be used only if the paint is sufficiently intact (with all layers present to permit an XRF measurement). However, if a previous inspection indicated that the deteriorated paint contains lead below the regulatory limit, it need not be sampled again. If the inspection was deficient, then repeated testing should be done.

Defective paint on furniture should be sampled in owner-occupied dwellings, since owners can resolve lead-based paint hazards on furniture. In landlord-owned dwellings, defective paint on furniture owned by the resident need not be sampled, since the furniture is not owned by the landlord. However, the risk assessor should recommend to the occupant that such paint be tested; responsibility for repair or removal rests with the occupant, not the landlord (unless the dwelling is furnished by the landlord). It may be in both the landlord's and the occupant's interests to go ahead and test the furniture paint during the first visit so that a clear determination of hazard can be made. If the furniture is not sampled, the risk assessor should note any deteriorated paint on furniture in the final report.

To collect a paint chip sample, remove *all* the layers of paint, but none of the substrate. This is ordinarily a relatively simple matter for defective paint, although it is important to collect *all* layers of paint from a sample location, whether they are all peeling or not. Do not sample only the peeling layers. For those layers that are still intact, a heat gun is recommended to soften the paint for easy removal. All dust or chips generated by the sampling process must be cleaned up.

It is usually not possible to report lab results in mg/cm^2 for defective paint, since the surface area sampled cannot be measured exactly due to paint deterioration. A complete protocol for sampling both intact and defective paint can be found in the HUD guidelines.

Intact paint in good condition should not be sampled as part of a risk assessment, since intact paint does not pose an immediate hazard. One exception concerns painted surfaces that will be disturbed as part of renovation or repair work. If the intact paint is expected to be disturbed in the near future due to repair and renovation work, the paint on those surfaces should be analyzed, even if it is presently in good condition. Intact paint on friction surfaces does not need to be sampled, since any lead dust that is being produced will be identified through dust sampling.

As a rule of thumb, no more than three to five deteriorated painted surfaces are sampled for most risk assessments. If more surfaces need to be sampled, the owner should be informed that it may be more cost-effective to complete a full lead-based paint inspection at the same time as the risk assessment.

Paint Samples from Chewable Surfaces

Only "chewable" painted surfaces that display evidence of teething, chewing, or mouthing constitute an immediate lead-based paint hazard that should be evaluated as part of the risk assessment process. If paint inspection has not been completed previously, or is not being completed as part of a concurrent lead-based paint inspection/risk assessment, or was not adequate, chewable surfaces can be analyzed by XRF. The inspector can also collect paint chip samples from surfaces that display teeth marks on protruding surfaces or from surfaces that the occupant reports are frequently mouthed by children. If collecting a paint chip, remove all the layers of the paint, but none of the substrate. Intact paint on chewable surfaces should not be sampled as part of a risk assessment. Limited cleanup with wet wipes and repainting should be done following any destructive paint chip sampling effort. Chewable surfaces include protruding accessible surfaces, such as window sills, stair toe boards, stair spindles, tables, furniture, and so on.

LEAD DUST SAMPLING

Due to the varied nature of housing construction, it will be necessary for the risk assessor to exercise professional judgment regarding the number of dust samples to be collected for a given residence. At a minimum, dust samples should be collected from the following rooms:

- Entryway
- Child's principal play area (usually the TV room, living room, or dining room)
- Two children's bedrooms (if only one child's bedroom is present, sample the child's room)
- Kitchen or bathroom

In addition, dust samples should be taken from the following locations in each room:

- Floors near friction or impact surfaces or deteriorated paint
- Window sills (from window most frequently opened)
- Window wells (from window most frequently opened)
- Cabinets with deteriorated paint housing dishes, toothbrushes, eating utensils, etc.

The risk assessor can conduct either composite or single-surface sampling. Composite sampling can reduce the total number of samples analyzed, but information on individual rooms is lost as a result of the compositing process. If single-surface sampling is conducted, additional samples are required, but the room-specific information generated may help focus cleanup efforts on particular rooms or areas, depending on results.

The complete field sampling and analytical method for collecting dust samples is contained in the HUD guidelines. Readers should note that these guidelines recommend the following precautions when conducting dust sampling:

- A standard wiping or vacuuming motion should be used
- Only certain brands of wipes should be used (unless equivalence can be demonstrated through blind dust spike sample analysis)
- Whatman filters and thick diaper wipes should not be used (Whatman filters are not sufficiently durable and thick diaper wipes cannot be digested by routine lab analysis)
- Field spiked wipe samples should be included in the sample stream in a double-blind fashion to ascertain the efficiency of the laboratory digestion procedure (this is necessary until laboratory proficiency testing programs include wipe sampling media used in the field)
- Hard-shelled containers (not baggies) should be used to contain wipe samples for transport from the sampling site to the lab, since the container should be rinsed quantitatively in the laboratory to recover all lead on the wipe.

Composite Dust Sampling

Composite dust samples should be collected using the following rules for combining subsamples:

- Four composite dust samples (one for hard floors, one for carpeted floors, one for interior window sills, one for window wells) should be collected. If all floors are carpeted or if all floors are smooth, a minimum of three composite samples is adequate. A window well (sometimes called the exterior window sill) is the part of the window in which the sash sits when closed. An interior window sill (sometimes called the stool) is the part of the window ledge facing the interior of the room.

- Composite dust samples can consist of either wipe or vacuum samples. The vacuum method should be used on carpeted surfaces. Wipe sampling should be used on hard surfaces. Because each of these methods has varying recovery rates, it will be necessary for the risk assessor to reference the appropriate EPA or HUD dust standard for each respective method.

- One composite sample should be collected from no more than four window wells: one in the principal play area, one in the kitchen, one in the youngest child's bedroom (older than 6 months), and one in the next oldest child's bedroom.

- One composite sample should be collected from no more than four interior window sills: one in the principal play area, one in the kitchen, one in the youngest child's bedroom (older than 6 months), and one in the next oldest child's bedroom.

- One composite sample should be collected from no more than four carpeted floors, if present: one in the principal play area, one in the kitchen, one in the youngest child's bedroom (older than 6 months), and one in the next oldest child's bedroom.

- One composite sample should be collected from no more than four smooth floors, if present: one in the principal play area, one in the kitchen, one in the youngest child's bedroom (older than 6 months), and one in the next oldest child's bedroom.

- Separate composite samples should be collected from carpeted and hard surfaces. In other words, a single composite sample cannot be collected from both carpeted and hard floors.

- Separate composite samples are required from floors, window sills, and window wells. In other words, it is not permissible to combine subsamples from floors, sills, and wells into one sample.

- Floor surface areas sampled in each room should be approximately the same size (nominally 1 ft^2 [929 cm^2]). Interior window sill and exterior well sampling size are dependent on window characteristics but should also be similar from room to room, if possible. In other words, the surface sampling area should not be skewed so that one room is over-sampled.

- If wipe samples are being composited, a separate wipe should be used for each spot sampled. Do not use the same wipe to sample two different spots. Do not insert more than four different wipes into a single container for a composite sample.

Single-Surface Sampling

A minimum of eight single-surface samples should be obtained from the following locations:

- One from the floor of the child's principal play area (TV room or living room)
- One from the interior window sill of the most frequently opened window in the child's principal play area
- One from the floor of the kitchen
- One from the window well of the kitchen window (if inaccessible, a window sill sample should be collected)
- One from the floor of the bedroom of the youngest child (older than 6 months)
- One from the interior window sill of the bedroom of the youngest child (older than 6 months)
- One from the floor of the bedroom of the next-oldest child (or adult's bedroom if there is only one child's bedroom)
- One from the window well of the bedroom of the next-oldest child, or adult's bedroom if there is only one child's bedroom (if inaccessible, a window sill sample should be collected). At least one window well sample should be collected in each unit.

This protocol provides for a mix of samples from floors, window sills, and window wells, with a preponderance of samples collected from floors more frequently contacted by children.

Additional Samples

The risk assessor should exercise professional judgment to determine if additional dust samples should be collected from other areas frequented by the child, or if certain samples should be substituted for others. For example, if certain windows are never opened, it may make little sense to sample the window well, since there is little potential for exposure. Similarly, if there are areas not identified above where there is visible accumulation of dust *and* obvious exposure to the child, then deviations from the sampling locations identified above are permissible. A tabletop in the child's play area that is dusty is an example of a surface that should be sampled in lieu of (or in addition to) the samples identified above.

If the owner requires statistically significant results with quantified confidence intervals, then an increased number of samples collected from surfaces selected randomly may be required.

Forms 7-2 and 7-3 respectively can be used for recording information on single-surface and composite dust sampling, respectively.

Common Areas (Multifamily Housing Only)

The number and locations of dust samples to be obtained from common areas in a multifamily building is determined by the type and number of common areas in the building.

FORM 7-2. Field Sampling Form For Dust (single surface)

Name of risk assessor ____________________

Name of property owner ____________________

Property address ____________________ Apt. No. ________

Dwelling selection protocol ______ All dwellings ______ Targeted ______ Worst-Case ______ Random

Target dwelling criteria (check all that apply)

______ Code violations

______ Judged to be in poor condition

______ Presence of two or more children between ages of 6 months and 6 years

______ Serves as day-care facility

______ Recently prepared for reoccupancy

Sample Number	**Room** (record name of room used by the owner or resident and and room number from floor plan sketch)	**Surface Type**	**Is Surface Smooth and Cleanable?**	**Dimensions of Sample Area** (inches × inches measured to nearest ⅛ inch)	**Area** (ft^2)	**Result of Lab Analysis** ($\mu g/ft^2$)
	Play room ______	Floor		___ × ___		
	Play room ______	Window sill (exterior or interior)		___ × ___		

	Kitchen ________	Floor		____ × ____		
	Kitchen ________	Window sill (exterior or interior)		____ × ____		
	Bedroom 1 ________	Floor		____ × ____		
	Bedroom 1 ________	Window sill (exterior or interior)		____ × ____		
	Bedroom 2 ________	Floor		____ × ____		
	Bedroom 2 ________	Window sill (exterior or interior)		____ × ____		
				____ × ____		
				____ × ____		

Total number of samples this page _______

Page _________ of _________

Date of sample collection ____/____/____ Date shipped to lab ____/____/____

Shipped by ________________________________ Received by ________________________________

(signature) (signature)

HUD standards: 200 μg/ft^2 (floors), 500 μg/ft^2 (interior window sills), 800 μg/ft^2 (exterior window sills)

FORM 7-3. Field Sampling Form For Dust (composite sampling)

Name of risk assessor ______________________________

Name of property owner ______________________________

Property address ______________________________ Apt. No. __________

Dwelling selection protocol ________ All dwellings ________ Targeted ________ Worst-Case ________ Random

Target dwelling criteria (check all that apply)

_______ Code violations

_______ Judged to be in poor condition

_______ Presence of two or more children between ages of 6 months and 6 years

_______ Serves as day-care facility

_______ Recently prepared for reoccupancy

Sample Number	**Record Name of Rooms Used by Owner or Resident to Be Included in Sample**	**Dimension of Surface Sampled in Each Room** (inches × inches measured to nearest ⅛ inch)	**Total Surface Area Sampled** (ft^2)	**Type of Surface Sampled**	**Is Surface Smooth and Cleanable?**	**Lab Result** ($\mu g/ft^2$)
	_______ _______ _______ _______	____ × ____ ____ × ____ ____ × ____ ____ × ____		Smooth floors		

	________ ________ ________ ________	____ × ____ ____ × ____ ____ × ____ ____ × ____		Carpeted floors		
	________ ________ ________ ________	____ × ____ ____ × ____ ____ × ____ ____ × ____		Interior window sills		
	________ ________ ________ ________	____ × ____ ____ × ____ ____ × ____ ____ × ____		Exterior window sills		

Total number of samples this page _______

Page _________ of _________

Date of sample collection ____/____/____ Date shipped to lab ____/____/____

Shipped by ________________________________ Received by ________________________________

(signature) (signature)

HUD standards: 200 μg/ft^2 (floors), 500 μg/ft^2 (interior window sills), 800 μg/ft^2 (exterior window sills)

• *Low-Rise and Mid-Rise Buildings* (up to and including three levels). Collect two dust wipe samples, one at the entry area floor and one from the first level landing floor in the common hall or stair connected to the dwelling. If there is a window present in the hallway that is opened and closed, substitute a window well or sill sample for one of the floor samples. Collect two additional samples from the top landing and stair tread. Again, if a window is present, substitute either the landing or the floor tread sample for a window sill or well sample.

• *High-Rise Buildings.* Collect two dust samples from the corridor of every fourth floor, one from the floors and the other from interior or exterior window sills. If there are no windows, collect both samples from the floors. Also, collect two dust samples from every fourth landing and stairwell, one from the landing floor and the other from the stair tread.

Corridor/Stairwell Inspection Locations

Record the conditions of all painted surfaces at all locations where samples are collected. For high-rise buildings, examine painted surfaces at levels from which samples are collected. For low and mid-rise buildings, inspect the entire hall/stair.

Dust Sampling in Other Public Spaces Frequented by Children

For spaces up to 2,000 square feet, collect dust samples as follows:

- *Floors:* Collect two samples from widely separated locations in "high traffic" areas regularly used or accessible to children.
- Window Wells/Sill: Collect two "worst case" samples, preferably from window wells.

  For spaces over 2,000 square feet:

- *Floors:* Collect one additional sample for each increment of 2,000 square feet.
- *Window Sills/Well:* Collect one additional sample for each additional increment of 2,000 square feet.
- *Management Office:* Collect one sample from the floor of the resident waiting area; two if area is more than 400 square feet.

SOIL SAMPLING

If present, bare soil should be sampled to determine if it presents a significant immediate hazard to children. In order to accomplish this, it will be necessary for the risk assessor to determine not only the concentration of lead in the soil, but also the use pattern (i.e., the frequency of contact, use of soil) for different soil locations and conditions. The risk assessor should also determine the extent of bare soil existing at a given dwelling. Bare soil should be sampled preferentially. Even if there is no bare soil, the risk assessor should collect soil samples to determine the importance of maintaining ground cover.

The following areas should be sampled for soil contamination:

- Outdoor play areas
- Soil near building foundation or along drip line
- Vegetable gardens, pet sleeping areas, bare pathways

A minimum of two composite samples per dwelling or building sampled are recommended: one from the child's principal play area, one from bare soil areas in the front or back yard (if present), and/or from the foundation drip line.

Samples may be collected using a coring tool to acquire the top half inch (1 cm) of soil. Alternatively, a stainless steel scoop may be used. Care should be taken to collect only the very top surface layer of soil. If soil is to be removed, it may be necessary to collect additional samples from a greater depth to determine the depth of soil removal.

Each composite sample should consist of soil subsamples collected from at least five distinct locations roughly equidistant from each other along an axis. For samples collected along the foundation drip line, subsamples should be collected at least 2 feet away from each other. For samples collected from bare areas or from the child's principal play area, samples should be collected at roughly equidistant points along each axis of an "x" shaped grid. Bare soil should be sampled.

If paint chips are present in the soil, they should be included. However, there should be no attempt to over-sample paint chips. The laboratory should be instructed to disaggregate paint chips by forcing them through the sieve. Use Form 7-4 for recording soil sampling information.

WATER SAMPLING

Water sampling can be conducted at the option of the building owner. Since there is an on-going lead-in-drinking-water program through EPA and local water authorities, and since corrective measures (corrosion-control water treatments) are often beyond the control of the owner, water sampling is not recommended as a part of routine risk assessments here. If water is suspected to be a major source of lead exposure in the locale under study, then samples can be collected and analyzed using standard EPA protocols.

ANALYTICAL PROCEDURES

All samples are to be analyzed according to standard methods. All laboratories analyzing lead in soil, dust, paint, and water should participate in EPA's National

FORM 7-4. Field Sampling Form for Soil (composite sampling only)

Name of risk assessor ______________________________

Name of property owner ______________________________

Property address ______________________________

Sample Number	Location	Bare or Covered	Lab Result (ppm)
	Building perimeter		
	Building perimeter		
	Play area 1 (describe) ____________		
	Play area 2 (describe) ____________		

Sketch soil sampling plot plan

Collect only the top ½" of soil

Total number of samples this page__________

Page _________ of __________

Date of sample collection ____/____/____ Date shipped to lab ____/____/____

Shipped by ___________________________ Received by ___________________________

(signature) (signature)

Lead Laboratory Accreditation Program and be accredited by an organization recognized by EPA (see Chapter 5).

Special Quality Control Procedures for Wipe Samples

It has been reported that the use of commercial wipe media results in low recovery rates in the laboratory due to inadequate digestion techniques.[3] Currently, there is no laboratory proficiency testing program that manufactures durable wipe material spiked with known amounts of lead dust. The Environmental Lead Proficiency Analytical Testing Program (ELAP) supplies Whatman filters spiked with known amounts of lead dust. However, Whatman filters are not sufficiently durable in the field and are not recommended for risk assessment or clearance sampling purposes. Instead, more durable wipe media, such as Little Ones Baby Wash Cloths from K mart or Little Ones Diaper Wipes, have been shown to exhibit recovery rates between 80 percent and 120 percent on a routine basis when spiked with lead dust.[4] The National Institute for Occupational Safety and Health (NIOSH) has also reported that Wash'n Dry wipes have acceptable recovery rates, although this has not been established in routine practice.[5] Other media may also have appropriate recovery rates, but should be evaluated on a product-by-product basis before use.

Because no national proficiency program examines laboratory performance of digestion procedures, it is necessary for risk assessors to insert blind spiked wipe samples with known amounts of lead dust deposited on the brand of commercial wipes being used at a frequency of 1 spiked wipe per 50 samples. These spiked samples are in addition to spiked samples prepared by the laboratory for its internal quality control/quality assurance program. Wipe samples should be spiked with lead dust in the range of 100–300 μg Pb/wipe, since 200 μg Pb/ft^2 is the region of interest and 1 square foot is the area usually wiped. Laboratories can usually prepare such spiked wipes upon request by risk assessors. The risk assessor should relabel (but not repackage) the spiked wipe samples so that the laboratory is as "blind" as possible to which samples are spiked samples and which samples are field samples. At present, this is the only means a risk assessor has of judging the performance of the laboratory's digestion procedure, since national proficiency testing program do not use the wipe media mentioned above.

Wet Chemical Field Test Kits

Although they demonstrate some promise for the future, wet chemical field test kits do not yet demonstrate sufficient reliability for routine use for analysis of defective paint, soil, or dust samples by risk assessors.[6,7] These kits are being evaluated in an extensive research study at the time of this writing.

Notes

1. C. S. Clark, R. L. Bornschein, P. Succop, S. S. Queltee, P. B. Hammond, and P. Peace. Condition and type of housing as an indicator of potential environmental lead exposure and pediatric blood levels. *Environ. Res.* 38:46–53.
2. National Decorating Products Association. *Paint Problem Solver,* 4th ed. St. Louis, 1993.
3. D. E. Jacobs. Evaluation of Digestion Procedures in Three Laboratories for Analysis of Spiked Lead Wipe Samples. Unpublished data, 1990.
4. Housing Environmental Services/Housing Authority Risk Retention Group. *Lead-Based Paint Risk Assessment Protocol.* Cambridge, MA, 1991.
5. National Institute for Occupational Safety and Health. *Evaluation of Field Methods for Lead Screening and Analysis.* American Industrial Hygiene Conference, May 1993.
6. Research Triangle Institute. *Evaluation of Lead Test Kits for Analysis of Paint, Soil, and Dust.* EPA Report 68-02-4550, 1991.
7. U.S. Environmental Protection Agency. *A Preliminary Evaluation of Commercially Available Lead-Based Paint Field Test Kits.* EPA Contract OD 4913 NAEX. Atlanta: Georgia Institute of Technology, 1991.

8

Risk Assessment, Part 3: Assessment of Various Housing Types

David Jacobs, C.I.H.

This chapter defines risk assessment protocols for four dwelling type/ownership combinations:

1. Homeowners with a single-family dwelling or small landlords operating fewer than 10 single-family rental units
2. Multifamily housing with informal maintenance staffs and only a single unit or a few units that are similar
3. Multifamily housing with formal maintenance staffs and numerous units that are similar (i.e., large landlords)
4. Landlords who own many different kinds of units

Table 8-1 summarizes the key elements of a risk assessment for each type of housing.

HOMEOWNERS AND SMALL LANDLORDS WITH FEWER THAN 10 UNITS

All dwellings in this category should be sampled individually because each one is different and the number of dwellings is small. Use Form 8-2 to ascertain the quality of the existing paint by direct observation.

Management Data

Complete Form 8-1 to assess the management capabilities of owners. An explanation of part 2 of the form follows.

Item (1), staffing. Smaller-scale multifamily housing is more likely to have a

TABLE 8-1. Risk Assessment Actions Required in Different Types of Similar Housing

Action Required	Multifamily Housing (10 or More Similar Units)	Multifamily Housing (2–9 Similar Units)	Owner-occupied Single-Family Housing or Small Landlords operating fewer than 10 rental units	Multifamily Units that are not similar (Large Landlords Operating 10 or More Rental Units)
Assess every unit	No	Yes	Yes	Yes[a]
Defective Paint sampling (if paint testing has not yet been completed)	Yes	Yes	Yes	Yes
Dust sampling	Yes	Yes	Yes	Yes
Soil sampling	Yes[b]	Yes[b]	Yes[b]	Yes
Water sampling	Optional	Optional	Optional (for extra fee)	No
Air sampling	No	No	No	No
Management system analysis	Yes	Yes	Not applicable	Yes
Maintenance work systems modified	Yes	Yes (if applicable)	Cleaning and repair practices modified	Yes
Housing condition and characteristics	Yes	Yes	Yes	Yes
Demographics & use patterns	Yes	Yes	Yes	Yes

[a]There may be occasions when all nonsimilar units need not be sampled individually.

[b]Reliable soil data collected for a neighborhood may be used in lieu of property-specific sampling.

simplified management structure. Indeed, the owner may also act as manager and maintenance worker. If there is a division of labor between owner and manager, or manager and maintenance worker, the risk assessor should attempt to determine who (if anyone) has the recognized authority to handle lead-based paint issues. One product of the risk assessment process is the clear designation of responsibility for dealing with lead-based paint issues.

Items (2), (3), and (4), lead-based paint poisoning prevention statement and previous lead-related efforts. Smaller-scale multifamily housing situations are also unlikely to exhibit a formal management statement on the issue. However, the risk assessor should encourage even small-scale landlords to adopt such policy statements as a planning exercise and as a demonstration of a good faith effort to

FORM 8-1. Management Data for Rental Dwellings

Part 1: Identifying Information

Identifying information:

Name of building or development ____

Number of buildings________

Number of individual dwelling units/building_________

Number of total dwelling units________

Date of construction _________ (if between 1960–1978, consider a screen risk assessment)

Date of substantial rehab, if any _______________

List of addresses of dwellings (attach list if more than 10 dwellings are present)

Dwelling No.	Address	No. Children Aged 0–6	Recent Code Violation Reported by Owner?	Chronic Maintenance Problem Reported by Owner?

Record number and locations of common child play areas (on-site playground, backyards, etc.)

Number _______

____________________ ____________________ ____________________

____________________ ____________________ ____________________

____________________ ____________________ ____________________

FORM 8-1. *Continued*

Part 2: Management Information

1. List names of individuals who have responsibility for lead-based paint. Include owner, property manager (if applicable), maintenance supervisor and staff (if applicable) and others. Include any training in lead hazard control work (inspector, supervisor, worker, etc.) that has been completed. Use additional pages, if necessary. This information will be needed to devise the risk management plan contained in the risk assessor's report.

Name	Position	Training Completed (if none, enter "None")
	Owner	
	Property Manager	
	Maintenance	

2. Is there a lead hazard control policy statement?
   _____ Yes _____ No (If yes, attach statement)
3. Have there been previous lead-based paint evaluations?
   _____ Yes _____ No (If yes, attach the report)
4. Has there been previous lead hazard control activity?
   _____ Yes _____ No (If yes, attach the report)
5. Describe dwelling turnover procedure, including typical cleaning, repainting, and repair activity.
   ____
   ____
   ____
   ____
6. Employee and Worker Safety Plan
   a. Is there an occupational safety and health plan for maintenance workers?
      _____ Yes _____ No (If yes, attach plan)

b. Are workers trained in lead hazard recognition?

_____ Yes _____ No   If yes, who performed the training? ____________________

c. Are workers involved in a hazard communication program?

_____ Yes _____ No

d. Are workers trained in proper use of respirators?

_____ Yes _____ No

e. Is there a medical surveillance program?

_____ Yes _____ No

7. Is there a HEPA Vacuum available?

_____ Yes _____ No

8. Are there any on-site licensed or unlicensed day-care facilities?

_____ Yes _____ No   If yes, give location ____________________________________

9. Planning for Resident Children with Elevated Blood Levels

a. Who would respond for the owner if a resident child with elevated blood lead level was identified?

____

b. Is there a plan to relocate such children?

_____ Yes _____ No   If yes, Where? ____________________________________

c. Do you (the owner) know if there ever has been a resident child with an elevated blood lead level?

_____ Yes _____ No _____ Unknown

10. Owner Inspections

a. Are there periodic inspections of all dwellings by the owner?

_____ Yes _____ No   If yes, how often? __________________________________

b. Is the paint condition assessed during these inspections?

_____ Yes _____ No

11. Have any of the dwellings have ever received a housing code violation notice?

_____ Yes _____ No _____ Unknown

If yes, describe code violation ____

12. If previously detected, unabated lead-based paint exists in the dwelling, have the residents been informed?

_____ Yes _____ No _____ Not applicable

reduce lead hazards. (See the section on large multifamily housing developments below.)

Item (5), turnover procedure. The risk assessor should determine what control measures, if any, are done when a unit is vacated and occupied by a new resident. Specifically, is the unit cleaned, how was it cleaned, and who decides and conducts any repair work that is needed?

Items (6a–e) and 7, employee health and safety plan. If the owner employs anyone as a maintenance worker or contracts for this work, the risk assessor should determine if the employee is trained to work with lead hazards safely, is aware of lead based paint hazards, knows how to use a respirator, currently owns a NIOSH/MSHA approved respirator, and has medical clearance to wear a respirator. If the owner does the work, are adequate protective measures in place? Are the workers trained to perform cleanup so that resident children are protected? Answers to these questions will inform the lead hazard control plan.

Item (8), day care facilities. Determine if any of the units are being used as a day care facility, based on the owner's knowledge. If so, that unit should be identified as such in the sampling effort.

Items (9a–c), Determine if the owner has made any plans to address resident children with elevated blood lead levels. The purpose of these questions is to help the owner begin the planning process necessary to respond to such a situation.

Items (10a, b) and (11), routine inspections. If the owner or manager conducts periodic housing quality inspections, determine whether or not those inspections examine the condition of painted surfaces. The risk assessor will often recommend that the owner or manager conduct periodic inspection to ensure that lead hazard control treatments retain their effectiveness. The presence of previous code violations may help focus attention on problem surfaces.

Maintenance Data

Use Form 8-2 to evaluate the visual condition of painted surfaces and the maintenance capabilities of the landlord (this does not apply to homeowners). An explanation of the form follows.

Items (1a–h), repainting policy. Do residents do all the repainting? Does the owner or manager supply the paint? Are the units repainted on a regular basis, and if so, who does it? How are paint and dust/chips removed? What protective measures are taken?

Item (2), preventive maintenance. Determine if a preventive maintenance program exists and, if so, how that program affects the condition of the painted surfaces.

Item (3), work orders. If a work order system exists, briefly describe it to determine how lead hazard control efforts can be integrated into existing practices.

Item (4), resident requests for maintenance work. The process by which

FORM 8-2. Maintenance Data for Rental Dwellings

Condition of Paint on Selected Surfaces				
Building Component	Paint Condition: Good, Fair, Poor, or Not Present (to be completed by Risk Assessor)	Deterioration Due to Friction or Impact	Deterioration Due to Moisture	Location of Painted Component with Visible Bite Marks
Building siding				
Exterior trim				
Exterior windows				
Exterior doors				
Railings				
Porch floors				
Other porch surfaces				
Interior doors				
Ceilings				
Walls				
Interior windows				
Interior floors				
Interior trim				
Stairways				
Radiator (or radiator cover)				
Kitchen cabinets				
Bathroom cabinets				
Other surfaces				

FORM 8-2. ***Continued***

If the overall condition of a component is similar throughout a dwelling, that condition should be recorded. If a component in a couple of locations is in poor condition, but the overall condition is good or fair, the specific sites of the badly deteriorated paint should be noted. The specific locations of any component with bite marks should be recorded.

1. Painting Frequency and Methods

   a. How often is painting completed? every ______ years

   b. Is painting completed upon vacancy, if necessary

      _____ Yes _____ No

   c. Who does the painting? ______ Property Owner ______ Residents

      (If residents, skip to Q.2)

   d. Is painting accompanied by scraping, sanding, or paint removal?

      _____ Yes _____ No

   e. How are paint dust/chips cleaned up? (check one)

      _____ Sweeping _____ Vacuum _____ Mopping _____ HEPA/wet wash/HEPA

   f. Is the work area sealed off during painting?

      _____ Yes _____ No

   g. Is furniture removed from the work area?

      _____ Yes _____ No

   h. If no, is furniture covered during work with plastic?

      _____ Yes _____ No

2. Is there a preventive maintenance program?

   _____ Yes _____ No

3. Describe work order system (if applicable, attach copy of work order form).

   ____

   ____

   ____

4. How are resident complaints received and addressed? How are requests prioritized? If formal work orders are issued, is the presence or potential presence of lead-based paint considered in the work instructions?

   ____

   ____

5. Record location of dwellings recently prepared for reoccupancy.

   ____________________ ____________________ ____________________

maintenance work is accomplished should be examined to determine how lead hazard control efforts can be integrated into existing practices. For example, the person responding to requests for maintenance services may have to know which requests may result in disturbance of lead-based paint. Work order forms (if they exist) may be modified to include special cleanup procedures or respiratory protection. The form should also include location of all units recently prepared for reoccupancy.

Occupant Questionnaire

For multifamily housing with nine or fewer units, it is feasible to use a questionnaire that will enable the risk assessor to focus attention on specific areas and units of concern. The questionnaire should be administered to all families in order to focus the sampling process on high risk areas. The questionnaire shown as Form 8-3 has been used in private housing.[1] Its use in large multifamily projects and for landlords with more than ten properties is not feasible.

TEN OR MORE SIMILAR UNITS (MULTIFAMILY HOUSING)

Buildings or developments with 10 or more similar units need not have every single unit assessed. Instead, a sample of dwellings can be selected to characterize the hazard for the entire population of similar dwellings.

For the purposes of risk assessment, "similar" means those dwelling units that were built at the same time, have a common maintenance and management history, have a common painting history, and are of similar construction. Similar units need not be contained in a single housing development or in a single building to meet this definition.

Random and Targeted Sampling

There are two ways of conducting sampling of many similar dwellings: random sampling and targeted sampling. The risk assessment protocol described here uses a flexible targeted sampling scheme to minimize the expense involved and maximize the number of units that can be assessed for the lowest cost. The protocol assumes that if units selected under a targeted sampling scheme are free of lead hazards, it is highly probable that the rest of the units are also free of lead hazards. While worst-case sampling has been widely used in the past with success,[2] it is not possible to quantify the degree of certainty associated with the findings. If the owner requires random sampling and a 95 percent or some other quantified degree of confidence, more units will need to be randomly selected and sampled.

FORM 8-3. Resident Questionnaire (to be completed by risk assessor via interview with resident)

Children/Children's Habits

1. (a) Do you have any children that live in your home? Yes ______ No ______
   (b) If yes, how many? _____ Ages? _____ _____ _____ _____ _____
   (c) Record blood lead levels, if known. _____ _____ _____ _____ _____

If no children, skip to Q.5.

2. Locate the rooms/areas where each child sleeps, eats, and plays.

Name of Child	Location of Bedroom	Location of All Rooms Where Child Eats	Primary Location Where Child Plays *Indoors*	Primary Location Where Child Plays *Outdoors*

3. Where are toys stored/kept? ________________________
4. Is there any visible evidence of chewed or peeling paint on the woodwork, furniture, or toys? Yes ______ No ______

Family Use Patterns

5. Which entrances are used most frequently? ________________
6. Which windows are opened most frequently? ________________
7. Do you use window air conditioners? If yes, where? ________________
   (Condensation often causes paint deterioration)
8 (a) Do any household members engage in gardening? Yes ______ No ______
   (b) Record the location of any vegetable garden. ________________
   (c) Are you planning any landscaping activities that will remove grass or ground covering? Yes ______ No ______
9. (a) How often is the household cleaned? ________________
   (b) What cleaning methods do you use? ________________
10. (a) Did you recently complete any building renovations? Yes ______ No ______
    (b) If yes, where? ________________
    (c) Was building debris stored in the yard? If yes, Where? ________________
11. Are you planning any building renovations? Where? ________________
12. (a) Do any household members work in a lead-related industry? Yes ______ No ______
    (b) If yes, where are dirty work clothes placed and cleaned? ________________

Number of Units to Be Sampled

The sampling schedule shown in Table 8-2 assumes that the units selected for sampling are targeted units (worst-case units), as defined below. The risk assessor should exercise professional judgment to ensure that the units actually selected are in fact worst-case units. If there is reasonable doubt that this is not the case (for example, if it appears the targeted units are being concealed or the necessary information is unavailable), then the risk assessor should resort to random sampling of a larger number of units. The results of random sampling can then be compared to targeted sampling to determine if the worst-case units are in fact worst-case. The targeted units should exhibit higher (or at least similar) lead dust levels. If the worst-case units exhibit lower lead dust levels, then the risk assessor should conclude that a targeted sample was not identified properly.

Selection of Targeted Units

Dwelling units should be selected so that they meet as many of the criteria listed below as possible. The purpose of the following criteria is to provide general guidance on factors to consider in defining a worst-case study population. However, it is not necessary for two different risk assessors to select exactly the same worst-case units in an apartment building, since there may be many that meet the following criteria.

If there are a number of units that all meet the same criteria, then the units with the largest number of children under the age of 6 should be selected. Children tend to increase wear and tear on painted surfaces, meaning that there is more likely to be a lead dust hazard in children's homes than in homes that are presently childless. If possible, at least one unit should have been recently prepared for reoccupancy, since repainting and other repairs are often conducted during vacancy, possibly creating a lead dust hazard. This unit should be sampled even if it is currently unoccupied.

TABLE 8-2. Number of Units to be Sampled in Multifamily Housing with Ten or More Similar Units

Number of Similar Units	Number of Units to Sample
10–75	10
76–125	17
126–175	19
176–225	20
226–300	21
301–400	22
401–500	23
500+	24 + 1 unit for each additional increment of 50 units or less

However, the risk assessor should not sample only those dwellings that have recently been cleaned and repainted, since this would not be an accurate representation of the conditions in occupied units in the rest of the development or building. The risk assessor should document which criteria were used to designate the unit as a targeted unit on the field sampling form.

In order of preference, targeted units should meet the following criteria:

1. Units that currently house children with elevated blood lead levels (20 μg/dl or a persistent 15 μg/dl). These units should be investigated using an expanded protocol, but only after consultation with the local health agency to avoid duplication of effort. If risk assessments of these units are to be completed, they cannot count toward the minimum number of targeted units to be sampled (i.e., each and every unit housing a child with an elevated blood lead level should be sampled, unless it has been fully investigated previously by local authorities). Insurance companies may want their risk assessors to conduct sampling in every unit containing an elevated blood lead level (EBL) child (if known). Since blood lead levels can be considered to be confidential medical information, a landlord may not know whether or not such children are present in the development at the time of sampling. However, a risk assessor should not ignore information on the prevalence of children with elevated blood lead levels at any given property if it is available, since a different method of investigation may be necessary. Failure to request such information would mean that the risk assessor may miss a unit that should have been investigated.
2. Units that contain two or more young children under 6 years of age but older than 6 months (the more young children the more "worst-case" the unit can be considered). If the landlord is uncertain about the number and ages of occupant children in each unit, a census should be taken by the owner.
3. Units with housing or building code violations within the past year.
4. If possible, one unit recently prepared for reoccupancy within the past three months should be selected to assess the quality of cleaning measures upon unit turnover.
5. Units that the owner believes are in the poorest condition (if the owner does not know, then use the other criteria or select units randomly).
6. If additional units are required to meet the minimum sampling number specified in Table 8-3, the risk assessor should select them randomly.

Management and Maintenance Data

Risk assessors should use Form 8-1 to evaluate management capabilities with regard to lead-based paint. Form 8-2 should be used to determine how lead-based paint hazard control can be integrated into the existing maintenance operations. The risk assessor should briefly explain each of the forms to be completed by the

TABLE 8-3. Example of Targeted Unit Selection

A risk assessor is hired to conduct a risk assessment for 20 units owned by a single landlord. Fifteen of these units are of similar construction and were acquired all at once. One of these 15 reportedly contained a child with an elevated blood lead level that has not been evaluated by local authorities, according to the health department and the landlord. The other 5 were acquired from different owners over the years with little previous rehab work and markedly different construction styles.

In this case, the risk assessor would conduct sampling in each of the 5 different units and 5 of the similar units (from Table 7.4, 15 units requires that 5 units be sampled), plus one more unit with the EBL child, for a total of 11 units to be sampled (5 different units + 5 worst-case units + 1 EBL child unit = 11 units). The EBL unit would be investigated using protocols specified by local authorities.

From the 15 similar units, the landlord has provided the following information:

6 units have 3 children under age 6
3 units have 2 children under age 6
5 units have 1 child under age 6
1 unit is vacant and has recently been prepared for reoccupancy

Of the 6 units with 3 children under 6 years of age, two had code violations. One of the units with 2 children has a history of chronic maintenance problems and is thought by the owner to be in relatively poor condition.

Based on this information, the risk assessor would select the following units as worst-case units:

2 units with code violations and 3 children under 6 years of age
1 unit prepared for reoccupancy
1 unit with the history of chronic maintenance problems

This yields 4 units. The final, fifth, unit to be selected would be one of the remaining 4 units with as many children as possible, or that exhibited one of the other criteria. If no such units exist, then the final unit can be selected randomly.

building owner or manager to make sure that the type and scope of information requested is understood.

LANDLORDS OF LARGE DEVELOPMENTS WITH DIFFERENT HOUSING UNITS

Current evidence suggests that risks in different dwelling units vary greatly, while risks in similar dwelling units appear to follow a pattern within the building or development. This means that a much greater number of units (if not all units) will need to be examined to determine if there is a pattern of immediately available lead sources, which makes it necessary to sample each unique housing unit to properly target control efforts. However, there may be circumstances in which clear patterns emerge even in a population of different housing units before each and every unit is sampled. For example, a landlord may find that the first 25 of 100 different houses to be sampled all have high lead dust levels in the window wells but nowhere else. In such a case, a landlord may conclude that all window wells are contaminated and should be cleaned in all 100 units and that it is cheaper to go ahead and clean all the window wells rather than test the rest of the units before cleaning. Furthermore, a landlord may decide it is cheaper to adopt a set of standard interim control measures recommended by a risk assessor and apply them to all units without first completing sampling in each unit.

Following this treatment, however, a qualified risk assessor should conduct clearance sampling and a risk assessment in each unit to ascertain the presence or absence of any other lead-based paint hazards that were not addressed by the standard interim control measures. In short, *all* units in this category should receive either a preliminary risk assessment and clearance testing or a followup risk assessment and clearance testing after a set of standard interim controls has been implemented. No unit should remain unexamined.

Additionally, it will not be necessary for the risk assessor to gather separate maintenance and management data for each unit, since in most cases these data will be the same (that is, the maintenance system will probably be the same for each unit under the control of the same landlord).

Landlords with fewer than 10 different units should have environmental sampling performed in each one, since the population is too small to see a pattern by selecting a sample of units. Landlords with more than 10 different units should select worst-case units as previously described. Use of the resident questionnaire is not feasible for most large landlords. For landlords with more than 10 different units, sample at least 10 worst-case units or 25 percent of the total number of units, whichever is greater. The risk assessor should evaluate the sampling results to determine if a clear pattern of some lead-based paint hazards can be discerned. If no clear pattern emerges, another 25 percent should be sampled and evaluated for hazard commonality, and so on. If a clear pattern of type and location of hazard can be identified, the landlord should determine if it will be more inexpensive to treat the homes that were not sampled than to sample them and proceed accordingly. Following this treatment, however, a certificate of lead safety for each different dwelling can be completed only after a risk assessment determines that no immediate lead hazards are present in each different dwelling. In other words, each different dwelling that gets a certificate should be examined at least once by a licensed risk assessor.

Notes

1. Adapted from J. Keck. Sample Risk Assessment Questionnaire for Private Single-Family Homes (unpublished). Baltimore: Leadtec Services, 1993.
2. *Occupational Exposure Sampling Strategy Manual.* NIOSH Publication No. 77-173. Cincinnati, 1977.

9

Risk Assessment, Part 4: Evaluating and Reporting of Findings

David Jacobs, C.I.H.

EVALUATION OF FINDINGS

The most challenging part of the risk assessor's task is to weave together management, maintenance, and environmental sampling data into a workable, practical, and affordable plan that corrects immediate hazards and places the property owner in an informed position to make certain that if and when new hazards appear, they can be controlled. Such an evaluation cannot occur until all test results, forms, and questionnaires have been completed.

Management

Except for the case of full removal of all lead-based paint, some type of on-going management of lead hazards is required. Concretely, that means assigning someone responsibility for managing the various aspects of a lead hazard control program. Assigning clear responsibility to a single individual is especially important for multiple owners and property management companies. If the landlord holds many properties, the owner (with input from the risk assessor) should determine which employee is best positioned to oversee training, management, periodic surveillance, response to resident reports of either defective lead-based paint or resident children with elevated blood lead levels, controlled maintenance and repair work, and so on. Often, this is someone in authority who has received previous training and who has demonstrated a concern about the issue. This person should be named in the written lead-based paint management program. This document need not be complex or even very detailed (see Form 9-1). It simply documents the landlord's recognition of the problem and the desire to do something about it,

FORM 9-1. Example of a Lead Hazard Reduction Policy Statement

XYZ Property Management Company is committed to controlling lead-based paint hazards in all its dwellings. ________________ (name), ________________ (position or job title), has my authority to direct all activities associated with lead hazard control, including directing training, issuing special work orders, informing residents, responding to cases of children with elevated blood lead levels, correcting lead-based paint hazards on an emergency repair basis, and any other efforts as may be appropriate. The company's plan to control such hazards is detailed in a risk assessment report and lead hazard control plan.

________________________ (Signed) ______________ (Date)
(Owner)

________________________ (Signed) ______________ (Date)
(Lead Hazard Reduction Program Manager)

and it authorizes someone to carry out a lead hazard control plan. The risk assessor should always recommend that the responsible individual acquire training.

On the other hand, if the landlord is a small one with no staff, or if the owner is also the occupant, there is clearly little point in naming an individual or adopting a formal policy statement. However, the risk assessor should always seek to educate the owner about the need to include lead hazard recognition in all other owner duties.

The unit turnover process should be reviewed to determine if work practices and cleaning efforts require modification. The risk assessor should determine what types of wet cleaning and repainting efforts can be achieved safely by the owner. The environmental data gathered from the unit recently prepared for reoccupancy should be examined to determine if these activities are making the most of a vacant unit situation, since hazard control measures are often much easier and cheaper to complete in vacant units.

The risk assessor should also determine if the owner's occupational safety and health program is adequate if workers will be repairing or otherwise disturbing surfaces with lead-based paint on a routine basis. Training is essential for such individuals to ensure that they are protected and that they do not inadvertently create lead hazards in the course of their duties in occupied units. The risk assessor should determine if respirator usage (and a respirator program), medical surveillance programs, and equipment purchases or borrowing arrangements (notably a high efficiency particle air, or HEPA, vacuum) are needed.

The risk assessor should help the owner determine what can be done quickly if a child with an elevated blood lead level appears. For example, can arrangements be made to secure a temporary lead-safe unit, either by coordinating with non-profit housing agencies or public housing authorities or by working with other property owners? At a minimum, the owner should know where alternative housing can be found on a rapid response basis.

Some landlords perform periodic general housing quality inspections, either on turnover or every year or so. The risk assessor should help the landlord find a way to include a closer evaluation of suspected or known lead-based paint condition during these routine inspections. This is best done by comparing the risk assessor's evaluation of the paint condition with that offered by the owner's representative.

Maintenance

In reviewing the information supplied on Forms 8-1 and 8-2, the risk assessor should determine if maintenance is adequate. Specifically, repainting should be performed at least every 5 years, more frequently if the condition of the paint appears to be poor.

If repainting is performed by the usual scraping and sanding, the landlord should be encouraged to use tri-sodium phosphate detergent (TSP) or other deglossing agent to prepare the surface, and/or change to wet scraping and sanding, followed by appropriate cleaning. The risk assessor should determine how well the owner's representative judged the condition of the paint. The work order forms (if they exist) should be evaluated to determine if they contain instructions about working on known or suspected surfaces with lead-based paint. For example, do the work orders tell workers when respirators or special cleaning are needed?

The quality of the maintenance operation should also be determined by evaluating the prevalence of code violations, the condition of paint, and the condition of the building as scored on Form 9-2. If the building is in "poor condition," or if there have been more than two code violations over the past two years, or if the condition of the paint is especially poor, then the risk assessor will conclude that maintenance is deficient and that lead-based paint hazards may not be adequately managed. Such a situation requires a more frequent reevaluation schedule (unless full removal is completed). Therefore, if the standard reevaluation schedule (SRS; see also the later section in this chapter) for the particular form of interim control or abatement selected is one year for well-maintained buildings, it could be every six months in buildings that are poorly-maintained. If the owner completes all necessary repairs, the SRS can be extended. Examples of standard reevaluation schedules are given in Table 9-1. Depending on the local jurisdiction, they may or may not be required.

Defective Paint

Results of Laboratory Analysis

All defective painted surfaces should be identified and listed separately in the final report, along with the laboratory results. Form 9-3 can be used for this purpose. Use Form 9-4 if composite paint chip sampling is being conducted. For composite

FORM 9-2. Building Condition Form

Condition	Yes	No
Roof missing parts of surfaces (tiles, boards, etc.)		
Roof has holes or large cracks		
Gutter or downspouts broken		
Chimney masonry cracked, bricks loose or missing, obviously out of plumb		
Exterior or interior walls have obvious large cracks or holes, requiring more than routine painting		
Exterior siding has missing boards or shingles		
Water stains on interior walls or ceilings		
Plaster walls deteriorated		
Two or more windows or doors broken, missing, or boarded up		
Porch or steps have major elements broken, missing, or boarded up		
Foundation has major cracks, missing material, structural leans, or visibly unsound		
Soil has numerous paint chips visible near foundation		
Total		
If the "Yes" column has four or more checks, the dwelling is considered to be in poor condition. Fewer than four checks in the "Yes" column means that the dwelling appears to be well maintained and the standard reevaluation schedule does not need to be revised.		

TABLE 9-1. Recommended Standard Reevaluation Schedules

Hazard Control Method	Standard Reevaluation Schedule[a]	Type of Reevaluation
No hazards identified[b]	2 years/4 years	Dust sampling, visual examination of suspect paint
Dust removal	1 year	Dust sampling, visual examination of suspect paint
Paint film stabilization	1 year	Dust sampling, visual examination of suspect paint
Friction and impact surface treatments	1 year	Dust sampling, visual examination of suspect paint
Encapsulation	Once after the first year, every three years thereafter	Dust sampling, visual examination of encapsulant integrity
Soil interim control	1 month initially, then 3 months, then annually	Visual examination
Soil abatement	20 years	Visual examination, soil sampling
Enclosure	10 years	Visual examination, dust testing
Removal of all lead-based paint	None	None
Building component replacement	None	None

[a]If the property remains lead-safe after the first two reevaluation periods, the schedule can be doubled thereafter. For example, if dust removal and paint film stabilization were completed, reevaluation by a certified risk assessor or inspector technician would occur after the first year, after the second year, and then after the fourth year, sixth year, and so on. If the reevaluation shows that hazards have reappeared, then annual reevaluations should be performed. The reevaluation period for soil abatement and enclosure should not be doubled due to the long time spans involved.

[b]If the risk assessment identifies no lead-based paint hazards, but lead-based paint (or suspect paint) remains in the dwelling, the next reevaluation should occur after 2 years. If no hazards are found again, reevaluation every 4 years is adequate. If an inspection indicates that no lead-based paint is present, no reevaluation is needed.

sampling, the standard must be divided by the number of subsamples included in the composite sample. For example, if there are five subsamples and the standard is 1 mg/cm^2, the criterion for the presence of lead-based paint is (1 mg/cm^2)/5 subsamples = 0.2 mg/cm^2.

If the concentration of lead in the paint is found to be less than 5,000 ppm by weight (also expressed as 0.5%, 5,000 mg/kg, or 5,000 μg/g), repair of the defective paint is not required (although it should still be recommended). Some states and jurisdictions use lower standards. If portable X-ray fluorescence (XRF) instruments are used for defective paint analysis, the appropriate standard is 1 mg/cm^2 (some locales also have lower paint standards as well that should be observed), although the risk assessor should exercise caution in determining whether the paint is too deteriorated to permit this kind of analysis. Specifically, paint chips *cannot* be analyzed by portable XRF, unless the chip is at least 3 inches by 3 inches

FORM 9-3. Field Sampling Form for Deteriorated Paint

Name of risk assessor ______________________________

Name of property owner______________________________

Property address ____________________________________ Apt. no. _______

Dwelling selection protocol ___ All dwellings ___ Targeted ____ Worst-Case___ Random

Target dwelling criteria (check all that apply)

____ Code violations

____ Judged to be in poor condition

____ Presence of two or more children between ages of 6 months and 6 years

____ Serves as day-care facility

____ Recently prepared for reoccupancy

Sample Number	Room	Building Component	Laboratory Result (ppm) or XRF Reading (mg/cm^2)
HUD Standard			5,000 ppm or 1 mg/cm^2

Sample *all* layers of paint, not just deteriorated paint layers.

Total number of samples this page _______

Page _______ of _______

Date of sample collection ____/____/____ Date shipped to lab ____/____/____

Shipped by ___________________________ Received by ___________________________

(signature) (signature)

Date results reported ____/____/____ Analyzed by ______________________________

Approved by ______________________________

FORM 9-4. Field Sampling Form for Deteriorated Paint (composite sampling)

Name of risk assessor ______________________________

Name of property owner ______________________________

Property address __________________________________ Apt. no. _______

Dwelling selection protocol ___ All dwellings ___ Targeted ____ Worst-Case ___ Random

Target dwelling criteria (check all that apply)

____ Code violations

____ Judged to be in poor condition

____ Presence of two or more children between ages of 6 months and 6 years

____ Serves as day-care facility

____ Recently prepared for reoccupancy

Composite Sample Number	Duplicate Subsample Number	Rooms Included in Sample	Component Sampled	Size of Subsample (cm × cm)	Lab Result (mg/cm^2)	Lab Result (ppm)
	________ ________ ________ ________	________ ________ ________ ________	________ ________ ________ ________	________ ________ ________ ________	________ ________ ________ ________	_____ _____ _____ _____
	________ ________ ________ ________	________ ________ ________ ________	________ ________ ________ ________	________ ________ ________ ________	________ ________ ________ ________	_____ _____ _____ _____
	________ ________ ________ ________	________ ________ ________ ________	________ ________ ________ ________	________ ________ ________ ________	________ ________ ________ ________	_____ _____ _____ _____
	________ ________ ________ ________	________ ________ ________ ________	________ ________ ________ ________	________ ________ ________ ________	________ ________ ________ ________	_____ _____ _____ _____
HUD Standard					1.0	5,000

Total number of samples this page _______

Page _______ of _______

Date of sample collection ____/____/____ Date shipped to lab ____/____/____

Shipped by __________________________ Received by __________________________

(signature) (signature)

and is intact. Otherwise, the surface area cannot be determined accurately. Although a method of grinding up paint chips and sprinkling them over a piece of paper has been recommended by one XRF manufacturer, equivalence with laboratory analytical methods has not been established.

Previous Paint Testing

If previous testing of lead-based paint has been completed, review the testing report for the characteristics identified below to determine if the information contained in the earlier report is reliable. Inspections completed years ago may not conform to current standards of care and may not have accounted for important sources of error, possibly resulting in an incorrect understanding of where lead-based paint is located. This will not necessarily result in repeated testing of all surfaces, although it may be necessary to augment the earlier testing.

The risk assessor should review the report of previous testing by answering the following questions:

1. Are there calibration check records for all XRF instruments used?
2. Were any "zero" readings made during the initial checking procedure on unpainted substrates supplied by the XRF manufacturer?
3. Were bare substrate effects accounted for and subtracted from the instrument apparent lead concentration where necessary?
4. Were all painted surfaces with visibly different painting histories tested?
5. Was an adequate number of units inspected?
6. Were the units properly randomized?

If the answer to any of the above questions is "No," the risk assessor should recommend that the owner treat those surfaces as potentially leaded above the regulatory limit, since results are uncertain. Therefore, if these "suspected" surfaces need to be disturbed during repair work or are in a deteriorated condition, they should be treated as if they contain lead-based paint. The owner should be encouraged to have all paint tested to make a final determination concerning whether any paint in the dwelling does in fact contain lead above the regulatory limit.

Dust

The risk assessor should evaluate the results of dust testing from worst-case and randomly selected units to ensure that the units labeled worst case are in fact that. If dust levels in worst-case units are similar to or higher than the randomly selected units, then the worst-case sample selection can be confirmed. If not, additional units may have to be sampled. If randomly selected units were not sampled, then of course no such comparison is possible.

Until the Environmental Protection Agency (EPA) releases its lead dust standard, the current Department of Housing and Urban Development (HUD) clearance standards should be used to determine if hazardous dust levels are present. Thus, if dust levels exceed any of the following levels, a lead-based paint hazard can be identified:

- 200 μg Pb/ft^2 on interior floors
- 500 μg Pb/ft^2 on interior window sills (window stools)
- 800 μg Pb/ft^2 on exterior window sills (window wells)

These standards were first developed using wipe sampling methods. Correlation with various vacuum sampling techniques has not yet been established, nor have correction factors for different surface dust sampling methods been calculated. Collection efficiencies of different sampling methods are known to vary.[1]

There are two ways of expressing dust lead levels, loading (μg/ft^2) and concentration by weight (typically ppm). Loading is usually thought to be more indicative of the total amount of lead in dust in the house, and therefore more representative of the immediate hazard. Additionally, it is clear that cleaning can greatly reduce loading, if performed properly. It is *not* clear that cleaning can reduce the concentration of lead in dust, since a cleaning method would need to preferentially remove lead from the dust to drive down concentration. Until further research is completed, it is best to rely on loading as the chief indicator of risk in all lead hazard control work and clearance procedures.

Two states have proposed different lead dust standards, one based on concentration by weight (Rhode Island, 1993) and another based on perceived cleanup feasibility (Minnesota, 1992). However, neither has yet been shown to be effective, enforceable, and practical on a broad basis. On the other hand, there is a significant body of experience suggesting that the HUD interim wipe standards cited above can be implemented broadly.[2,3]

For large data sets, the risk assessor should prepare summary statistical measures of lead dust on floors, window sills, and window wells where units are similar (arithmetic or geometric mean and standard deviation). Calculators and computer programs are readily available to perform such calculations. If the mean of the worst-case units exceeds the dust standards, a lead dust hazard has been identified for all dwelling units in the population. If the mean is less than the dust standard, a dust hazard probably does not exist. If it is greater than the standard, then a lead hazard has been identified for all units.

Statistical calculations are not required for small data sets (fewer than 10 dwellings). Such calculations are also unnecessary when a cursory review of the data show that most of the individual sample results for a particular type of surface are greater than the standard.

The data should be examined to determine if consistent patterns emerge. For example, do the window wells contain high levels, while floors and sills are low? Do common hallways have high levels on floors, while other floors are low? Such patterns will aid in the development of recommendations for focused, less expensive control measures.

Soil

Evaluation of soil sampling results are hampered by the lack of an accepted standard. The EPA currently has a Superfund cleanup standard of 500–1,000 ppm, although it does not have a residential soil lead standard. The State of Minnesota has a bare soil standard of 300 ppm. The EPA Three Cities study suggested that statistically significant reductions in blood lead levels do not occur until soil to be removed exceeds 2,000 ppm.[4] The mean blood lead decline was 0.9–1.2 μg/dl.

The risk assessor should evaluate both the condition and use of the soil in determining if a hazard is present. Until EPA promulgates its soil standard, the following rules of thumb may be used in assessing lead soil hazards:

- Only bare soil is immediately hazardous
- Play areas with bare soil that are frequented *often* by young children should not exceed 500 ppm
- Play areas with bare soil that are frequented only periodically by young children should not exceed 1,000 ppm.
- Other bare soil areas that could be tracked into or blown into the dwelling (e.g., the building perimeter), but are contacted only infrequently by young children should not exceed 2,000 ppm.

Corrective measures usually include some form of covering the soil with grass or sod, or making it inaccessible to children by planting thorny bushes.

Water

The current EPA action level standard for lead in drinking water is 15 ppb (15 μg/L) drawn as a 1 liter first flush after the water has remained in the pipe for at least 6 hours. If first draw tap water samples exceed 15 ppb Pb, the risk assessor should recommend that the homeowner contact the local water department to determine if corrosion control or other control measures are in the process of being implemented. Because control of lead in drinking water is often not related to treatments at the dwelling, collection of drinking water samples is not a routine part of risk assessments.

Other Lead Sources

If other lead sources are discovered in the dwelling, the risk assessor should contact the local health department or the local childhood lead poisoning prevention program for assistance in devising control strategies and assessing the degree of risk. If it appears that a parent or other occupant works in a lead industry and is bringing lead hazards into the house, the Occupational Safety and Health Administration (OSHA) should be notified, since the OSHA lead standard contains important provisions to prevent workers from "taking home" occupational lead dust.

REPORT

Site-Specific Risk Management Options

Once the nature, severity, and location of lead hazards are understood, the risk assessor should advise the owner on the wide range of measures that can be undertaken to control those hazards (see Table 9-2). These control measures range from various types of short-lived interim controls (everything from specialized cleaning to minor

TABLE 9-2. Main Hazard Control Options That Could Be Identified in Risk Assessments

Treatment Option	Dust on Floor	Dust on Windows	Paint on Doors	Paint on Windows	Paint on Floor, Walls	Paint on Trim	High Soil-Lead Levels
Dust removal	x	x	x	x	x	x	x
Paint film stabilization			x	x	x	x	
Friction reduction treatment	x	x		x		x	
Impact reduction treatments	x	x	x			x	
Planting grass	x						x
Planting sod	x						x
Paving the soil	x						x
Encapsulation					x	x	
Enclosure					x	x	
Paint removal by heat gun			x	x		x	
Paint removal by chemical			x	x	x	x	
Paint removal by contained abrasive			x	x	x		
Soil removal							x
Building component replacement			x	x	x	x	

Note: Column headings refer to lead-contaminated dust and deteriorated lead-based paint.

wet scraping and repainting) to more long-term abatement methods, such as building component replacement, covering, and paint removal. Generally speaking, the greater the longevity of the control method, the greater the initial expense.

Cost

Each owner will have different levels of funding available. Some will be able to make a long-term investment that will be more inexpensive in the long run and should add to the value of the property. Others will be unable to make such an investment and will opt for short-term measures that are more inexpensive in the short term but require more frequent monitoring and reevaluation. The risk assessor should endeavor to provide information to the owner to permit an informed decision on this complex issue. Costs for various treatments vary considerably from one locale to the next and are also subject to market conditions, making it difficult to provide firm cost figures here. However, the risk assessor should provide a rough estimate of cost for each control option based on local conditions.

Feasibility

In addition to cost, the risk assessor should decide which treatments are likely to be ineffective. Examples of ineffective control measures include the following:

Repainting an area of defective paint caused by moisture problems (leaky roof, poor vapor barrier, uncorrected plumbing problem, window air conditioner, etc.). The moisture problem should be corrected first.
Repainting or encapsulating an area subject to impact and friction.
Planting grass seed in high traffic areas.
Replacement of window sashes in frames that are severely deteriorated.
Cleaning of surfaces that are not sealed or made "cleanable" or failing to identify and correct a lead dust source.
Mulching or covering leaded soil in areas where pets tend to sleep.
Dry sweeping.
Washing horizontal surfaces without stabilizing chalking vertical painted surfaces.

Of course, the risk assessor should also emphasize the severe danger of using prohibited methods, such as torching, power sanding, and uncontained abrasive, sand, or water blasting of painted surfaces.

RISK ASSESSMENT REPORT FORMAT

The following format is recommended for risk assessment reports:

Part I: Identifying information

1. Risk assessor name and license no. (where applicable)
2. Owner name, address (if different than above), and phone number
3. Date

Part II: Results of management, maintenance, and environmental analysis

4. Management information—Form 8-1 or Form 8-3 (not required for homeowners)
5. Maintenance information—Form 8-2 or Form 8-4
6. Brief narrative description of unit selection process (not required if all units were sampled)
7. Dust sampling results—Form 7-1 or 7-2
8. Defective paint sampling results—Form 7-1
9. Soil sampling results—Form 7-4
10. Other sampling results (if applicable)
11. Building condition—Form 9-1
12. Analysis of previous XRF testing report (if applicable)
13. List of location and type of actual or suspected lead hazards

Part III: Management plan

14. Lead-based paint policy statement (not applicable for homeowners)
15. Name of individual in charge of lead-based paint hazard reduction program (not applicable for homeowners)
16. Recommended changes to work order system (not applicable for homeowners or landlords without work order systems)
17. Recommended interim control options and estimated costs
18. Recommended abatement options and estimated costs
19. Reevaluation schedule
20. Interim control/abatement option to be implemented in this property
21. Method of resident notification of results of risk assessment and lead hazard control program (not applicable for homeowners); should include discussion of how residents are to be educated about lead poisoning, *before* risk assessment results are released)
22. Signatures (risk assessor and owner) and date

A certificate of lead-safety (with an expiration date if all lead-based paint is not removed) may be provided by the risk assessor, inspector, or local enforcement agency following the implementation of the accepted interim control/abatement option and any associated clearance sampling.

STANDARD REEVALUATION SCHEDULES

Unless the owner selects and actually completes full lead-based paint removal, reevaluation will be required to ensure that lead hazards do not reappear. A reevaluation schedule is a program for the frequency of repeated reevaluation by a certified professional recommended for each type of control method. Some forms of abatement require no reevaluation at all, while others need it every 20 years, 5 years, 3 years, 1 year, or less. Depending on the results of the risk assessment, the standard reevaluation schedule (SRS) may be adjusted (see Chapter 5). Some jurisdictions may use an SRS for determining how long a certificate of lead-based compliance is valid. See Table 9-1 for a listing of the recommended SRS for each control method.

Those property owners who demonstrate that the lead hazards in their dwellings are being effectively controlled are rewarded by providing for reevaluations at less frequent intervals. For example, if leaded dust levels in a dwelling are below clearance levels and all painted surfaces are stabilized and, upon reevaluation a year later leaded dust levels are still satisfactory and all painted surfaces are still intact, owners can conduct their own monitoring for the next year or two. In these low-risk situations, less frequent independent reevaluations are appropriate.

Property owners with solid compliance records should be required to prove their compliance less and less frequently, thus releasing funds for more permanent abatement activities and high-risk situations. State and local governments should be encouraged to test such self-monitoring approaches, particularly when the property owner combines interim control measures with more permanent control of lead hazards.

In certain circumstances, the risk assessor should shorten the SRS. The following guidelines should be observed when making this adjustment:

- *Management:* If the property manager or owner demonstrates a clear lack of concern, is severely understaffed, refuses to train its workers or acquire the necessary equipment, refuses to develop a contingency plan to handle lead-poisoned children, or refuses to correct clear management deficiencies, the risk assessor should document such evidence and reduce the SRS by an appropriate amount (i.e., if the SRS is 1 year, the adjusted SRS for poor management may be 6 months). In some jurisdictions, risk assessors may have a legal duty to report owners who fail to correct lead hazards to local enforcement authorities.

- *Maintenance:* If the building is in poor condition (as determined on Form 7-1) or if the risk assessor has a documented reason to believe that maintenance of painted surfaces is likely to be deficient, the SRS should be reduced by an appropriate amount.

• *Preexisting Dust Lead Levels:* If preexisting dust lead levels exceed the current standard by more than a factor of 10 on floors and window sills and 100 on window wells, there is evidence that the dwelling is at increased risk of rapid reaccumulation. Therefore, if preexisting *mean* dust levels exceed the following limits, the SRS should be reduced by half:

2,000 μg Pb/ft^2 on floors
5,000 μg Pb/ft^2 on interior window sills
80,000 μg Pb/ft^2 on exterior window wells

Notes

1. D. E. Jacobs, W. B. Menrath, P. A. Succop, R. Cohen, C. S. Clark, R. L. Bornschein, and M. Farfel. *A Comparison of Five Sampling Methods for Settled Lead Dust: A Pilot Study.* Columbia, MD: National Center for Lead Safe Housing, June 15, 1993.
2. *Procedures for Abating Lead Containing Substances from Buildings.* COMAR 26.02.07, Title 26. Baltimore: Maryland Department of the Environment, 1988.
3. U.S. Department of Housing and Urban Development. *Lead-Based Paint: Interim Guidelines for Hazard Identification and Abatement in Public and Indian Housing.* Washington, DC, September 1990.
4. U.S. Environmental Protection Agency. *Review of the National Ambient Air Quality Standards for Lead.* EPA-450/2-89-011. Research Triangle Park, NC, 1989.

10

Managing Residential Lead-based Paint Hazards: Abatement

Keith W. Pokorny

The goal of lead-based paint abatement is obvious to all involved in the process: to remove the lead hazard from the immediate living environment of the dwelling occupants. To achieve this goal there are many steps to take, whether the abatement is performed in one room during a residential upgrade project or in a federally funded "Comprehensive Modernization" project at a multiunit housing development. In all cases, the abatement process must be regarded as a set of procedures involving a known potentially hazardous substance. Therefore the abatement process must follow basic industrial hygiene principals such as administrative controls, engineering controls, worker protection, and proper procedures coordinated to achieve the goals of the project.

This chapter begins by defining the goals of an abatement project, evaluating the available options, and finalizing the scope of work. Abatement options are discussed in detail, including the advantages and disadvantages of each method and the applicability of new technology to the lead abatement industry. Also discussed is the relationship of other trades involved in the abatement process and the need to carefully design an abatement strategy when approaching larger-scale projects. Related work items such as the installation of removed components and the painting of surfaces stripped of lead-based paint are part of the project and involve the integration of other trades such as carpenters, painters, electricians, and so on into the overall project schedule. The design of an abatement project, no matter what the size, is essential to a successful project, meeting the goals of eliminating the hazard and providing a lead-safe living environment.

GOALS OF A LEAD ABATEMENT PROJECT

The first step in the lead abatement project is to clearly define the goals of all parties involved. In residential, single family abatement projects, the resident of the home and/or the owner is generally the primary constituent. Lead abatement in these situations generally arises from an owner performing renovation activities and realizing that the existence of lead paint is a concern. While lead abatement regulations do not exist in most states, the owner may be required to follow specified procedures in those states that do have restrictions. The hiring of trained, licensed abatement workers is required in some states and proof of this may be required to receive compliance with the states' regulations. (In Massachusetts, an owner is required to present a copy of the paid invoice from a licensed lead abatement contractor and receive a "letter of compliance" from a third-party licensed lead inspector to officially be considered in compliance with the state's Lead Law). It is always recommended that the owner hire a contractor with a proven track record in performing lead abatement safely.

An owner must be aware of the hazards created when lead abatement is not properly performed. Dust that contains lead will contaminate adjacent rooms or areas close to the abatement process if a work area is not properly defined, contained, cleaned, and inspected. The owner must require the contractor to create a designated and identified work zone, by room or area. This work area must be properly isolated to prevent the contamination of adjacent rooms or areas. After abatement is completed and at the end of each work shift, the contractor must decontaminate the work area from generated lead dust. The final step is a thorough visual inspection to ensure proper cleaning was performed.

Other steps including the use of high phosphate soaps (check local regulations regarding the use of phosphates in your community) and high efficiency particle air (HEPA) vacuums are part of the abatement process. Retrieving dust wipe samples for work area clearance is recommended to verify adequacy of cleanup. These basic principles are used throughout the lead abatement industry and all are expanded upon in this chapter. The main goal in any abatement process is to remove the present or potential hazard. As stated later, the importance of a contract between the owner and contractor is essential to protect all parties involved in the abatement process. An abatement project is more likely to be successful if the principles explained here are followed.

Large-scale lead abatement projects typically involve many players with integrated roles and responsibilities. While the overall goal of the abatement project remains the same as a small-scale project, the roles and responsibilities of the parties involved present different challenges in ensuring a smooth abatement process that meets the original goals of the contract within its budgetary constraints.

Examples of owners involved in large-scale lead abatement projects range from public funded state and federal housing developments to privately owned property and essentially any form between the two. There are privately owned developments that receive public funding (federal, state, or local), foreclosed property owned or managed by banks or financial institutions, and property funded from all these. It is important to understand the funding source needs and requirements in any project, as this can directly affect the outcome of the final product. Budgetary constraints are perhaps the greatest challenge in designing an abatement strategy. These constraints will define abatement options and the overall goals of the abatement process itself. For this reason, an owner or a funding participant may hire a financial consultant knowledgeable in construction trades to assist in overseeing any decisions affecting the project budget, construction period, and payment issues. Rental income is based solely on occupancy, so the time-frame of a project is also first and foremost on the list of the owner and financial consultants and funding agencies.

The owner may also require the services of a project architect, environmental consultant, and various subconsultants to provide project planning, bidding, and contract administration. Their role is to coordinate the design activities and to specify to the contractor the methods and materials used in performing the work.

Public funding agencies will usually assign a project manager representing the agency to oversee activities for the entire project from start of design through completion of construction. The owner may have an in-house project manager or construction manager representing the owner in all aspects of the project.

The owner will enter into a contract with a contractor responsible for the performance of the abatement project. In many cases, lead abatement is only a part of a larger renovation or rehabilitation project. As is similar to the many layers defining an owner, the contractor is usually a general contractor who hires subtrades to perform part or all of the renovation. These subcontractors can include the lead abatement subcontractor, electricians, carpenters, window installation subcontractors, painters, and so on. Essentially, the contractor or group of contractors has entered into a contract with the owner to perform defined items of work for a specified cost. The contractor who won the bid for work has a project budget to perform the work, pay the labor, buy the specified materials, and cover fixed costs, and overhead, plus generate a profit. The goal of the contractor is to meet the contract requirements within the project budget and to the satisfaction of the owner.

As previously stated, the abatement project, not including the actual abatement and construction decisions, can become complicated because of the number of different parties involved and each party's goals and expectations for a successful completion. Understanding each party's role is essential. The abatement planner must constantly be reminded of this in every step of the abatement process, starting with the abatement design.

EVALUATING THE OPTIONS IN ABATEMENT PLANNING

Preliminary Design Activities

The first step in the design or planning of an abatement project is a complete review of the testing procedures performed to identify components with lead-based paint. The planner should carefully review the testing results to ensure the proper identification of components considered for abatement. Once test results are reviewed, a list should be compiled of all components included in the abatement process. For example:

Windows	All sashes, exterior sills
Baseboards	Kitchens only
Door casing/jamb	Front and rear entrance, pantry, misc. bedrooms
Doors	Front entry

To decide the appropriate abatement methods for each identified component, the planner must perform the following:

1. *Consult with the owner.* Review the goals of the abatement project. Is it the intent to remove the hazard altogether, or are there other options the owner is willing to consider?
2. *Review the budget.* Are there budgetary constraints that prohibit some abatement methods (e.g., component removal may not be an option due to the replacement costs, so paint removal or enclosure may be the best options)?
3. *Review the long term/short term.* It is the planner's responsibility to the owner to recommend the most cost-effective solution providing long-term hazard reduction or permanent removal of the hazard. Short-term solutions like encapsulating products or making lead paint intact in order to remain in place should be recognized for what they are, short-term solutions. Reviewing the overall use of the facility will also provide options or narrow procedures recommended. For example, a public housing authority (PHA) may require the project serve as a 10-year upgrade of the facility, thus requiring long-term solutions. Full-scale rehabilitation/renovation will also require long-term solutions such as the renovation of a facility into a daycare center.
4. *Review regulatory requirements/funding agency requirements.* As stated previously, there may exist federal and state regulations for performing lead abatement. Also the funding sources may have specific requirements based upon set policies or insurance requirements. Federally funded housing must comply with the guidelines and deadlines of HUD. Some state funding agencies have in-house policies regarding the types of abatement to be performed, such as

requiring component removal for long-term solutions. Banks and financial institutions may also have in-house policies. Any of these requirements may place restrictions on the abatement methods to be performed, types of waste generated, the training of workers performing the abatement, and other aspects of lead abatement.

Abatement Options

Although lead abatement is a relatively new field, there exists many tables, charts, and studies referring to the various abatement methods known to exist. These include comparing costs, long-term/short-term benefits, related work, and the advantages/disadvantages. While these charts and tables are useful, they should never be solely relied upon for abatement decisions and selection. As shown in the case studies presented at the end of this chapter, field conditions at each site will affect the outcome of the abatement decision process. It is strongly recommended that field conditions be thoroughly explored prior to evaluating the abatement options.

The following discussion reveals the advantages and disadvantages of various abatement options. It is important to be knowledgeable of the limitations of each method, since this will affect the abatement decisions and effectively serve the owner and resident. The following outline summarizes the available abatement methods:

1. Component removal/replacement
2. Paint removal
   a. Chemical paint removal—on-site
   b. Chemical paint removal—off-site
   c. Mechanical paint removal
   d. Manual paint removal—dry scraping
3. Enclosure
   a. "permanent"
   b. short-term
4. Encapsulate
   a. Product encapsulants
   b. Covering methods

Component Removal/Replacement

Removal of the component painted with lead-based paint is obviously the most effective solution to removing the lead hazard from the living environment. The other advantage to choosing this method is the resulting upgrade of the existing structure or component. Replacing old leaded window sashes and casing (Fig. 10-1) with new energy efficient windows serves three purposes: removal of the hazard, increased energy efficiency, and an upgrade of property. The combination of these

FIGURE 10-1. This lead-painted window sill posed an immediate risk of lead poisoning to residents. All window components and casing were removed and replaced.

advantages will increase the overall value of the property and hopefully extend the life of the structure.

Another advantage to component removal is that in many cases it is the easiest and quickest method of abatement, and involves the least aggressive types of cleaning methods. The removal and replacement of stair railings/balusters or molded trim components, for example, may generate little lead dust or paint chips if the paint is intact. Additionally, component removal avoids paint scraping or feathering of paint lines usually required during paint removal operations.

Two major reasons why component removal may not be the recommended abatement method are cost and workmanship quality. Replacement costs of windows range from $80 per window sash to $500. Prefabricated windows, proper glazing, special installation requirements, and inconsistent field conditions may increase the cost of replacement. Also, depending on the lead content and lead leachability from the removed substrate, the waste generated during the operation may be considered hazardous. On a large-scale project, the cost for hazardous waste disposal can greatly increase the overall cost if not properly controlled. (See the subsequent section on waste disposal.)

Similarly, replacement of decorative trim and intricate design woodwork can be expensive. The pleasing aesthetics of the building may be the first casualty of budgetary constraints. Lead abatement in a Victorian style home becomes obviously more difficult and expensive than that of a basic ranch-style home. During removal/replacement activities, work must proceed carefully. Proper installation of replacement materials requires skilled tradesmen to meet product warranties. Product and contractor warranties for the work must be detailed for the contractor in the project specifications and contract.

The time frame of a project may also be affected by the delivery of replacement materials and may result in cost overruns, contract disputes, and a delay in reoccupancy. Delays from suppliers, rejection of submitted product samples and cutting corners on the quality of replaced materials are a reality in the administration of the abatement project.

Cautionary notes: Preexisting construction conditions must be examined when specifying component removal/replacement. As explained in the case studies at the end of this chapter, what the original construction drawings or the as-built drawings show as existing may not exist in the field. This may be due to faulty construction, normal aging of a building, settling, or incorrect as-built drawings.

Another factor to be considered is that the lead abatement contractor may not necessarily employ skilled workers to perform finish carpentry, window installation, painting, and so on. Third-party oversight of the subcontractor or general contractor is recommended for the majority of component removal if the previous work of a lead abatement contractor cannot be viewed or referenced. The owner or planner must check references and specifically ask questions about the abatement contractor's limitations of workmanship or the ability to work with other trades.

The planner or owner may rely on the skills of an experienced architect to assist in the design of replacement processes. A project team consisting of an experienced environmental consultant with construction experience and architect with hazardous materials experience is highly recommended.

Paint Removal

There are many paint removal processes available in lead abatement including the use of chemical strippers, mechanical strippers, and manual scraping methods. While each method is different, requiring varying levels of skill, all methods have the distinct advantage similar to component removal: The primary result is (intended to be) the removal of the hazard permanently from the living environment.

Chemical Paint Removal

On-site paint removal with the use of chemical strippers can be very effective as an abatement method, and also cost-effective. There are a variety of paint removal products and manufacturers. The most common type of product is a caustic material applied to the substrate and later removed, taking the paint with it. Commonly termed *caustic paste,* these products are very effective with lead-based paint on a variety of surfaces. Some manufacturers supply several varieties of their product with differing degrees of strength for use on defined types of substrates. There is generally a three-step procedure to the use of caustic paste: (1) paste application to the painted component or substrate, (2) allow a settling period for the product to react with the paint (commonly termed the "dwell time"), (3) remove of the product from the surface (along with the paint) (see Fig. 10-2).

For many products, there is a fourth step—the neutralization procedure that returns the surface to the proper pH for repainting or finishing. Testing the surface for pH is required because caustic pastes are highly alkaline and leave a film to which new paint will not adhere.

Caustic pastes can be dangerous and require an additional level of worker protection. This usually includes a face shield and protective gloves to prevent skin burns caused by the alkalinity of the product. Also, the products require a level of skill and a familiarity with the proper use of each variety of product for successful paint removal. While this may be considered a disadvantage to this abatement method, the proper use of many of the available products results in the nearly complete removal of lead paint from the surface and eliminates the need for component removal and replacement. This is a major advantage when abating detailed woodwork and trim with intricate designs. Many owners prefer this method of abatement to retain the pleasing aesthetics of the architectural components.

Additionally, some caustic pastes are very effective on metal components such as steel stair towers, railings, and most steel components. A skilled abatement contractor will also be aware of the useful qualities of the alkalinity of the product. Once the paint is removed from a metal surface and the surface is neutralized as a final step, rusting of the surface becomes a problem. Within hours of neutralization, the surface will exhibit a rust haze, the beginning of surface rusting. If, however, the neutralization is delayed until just before painting of the metal, rust is prohibited. Since the metal surface in contact with the air remains at such a high pH before neutralization, oxidation of the surface does not occur or is delayed for several days.

As with any product, proper testing of its capabilities and effectiveness will ensure success in its use. For this reason the owner should insist on employing "test patches" on various substrates to be abated. This concept has been widely successful in identifying which surfaces should be abated with caustic paste and where the product is unsuccessful. To create a test patch, apply the caustic paste to a defined area on the substrate (e.g., a 1-square-foot area of a plaster wall or one tread and riser of a stairwell). Follow all manufacturer's instructions, and after the appropriate dwell time and removal of the caustic, determine if all of the visible paint has been removed. Keep in mind that a slight residue may exist, which should wash off during neutralization. Complete the test patch process by neutralizing and testing to see if the surface can be repainted or refinished to satisfactory results. For large-scale projects that will potentially use a large quantity of caustic paste, the manufacturer may provide the test patches for you at no additional cost.

A disadvantage to caustic pastes is their ineffectiveness if different types of paints have been applied. For example, if lead-based paint exists on a surface in

FIGURE 10-2. *Top:* Test patches of different formulations of caustic paste were applied to a steel column. *Bottom:* After one day the pastes were scraped off and the most suitable formulation for this component was apparent.

several layers but there is a layer of latex paint between two layers, some caustic pastes have been known to stop at the latex paint leaving a muddied residue that must be removed before a second coat is applied. In many cases, scrubbing of the surface at this point may increase the penetration of the caustic paste through the latex. Damage to the surface may occur (e.g., plaster walls), so caustic pastes may not be applicable to all surfaces. It must be noted that the effectiveness of each product may vary to the degree to which it is affected by latex paint, and the manufacturer should be consulted.

Any contractor or worker using caustic paste must be trained in the dangers of the product. Most caustic pastes have a high pH. This can cause severe burns to the skin and permanent damage if it gets in eyes, nose, or is ingested. Also, some strippers are in the form of powder that is mixed to form the paste. If a worker gets the powder on the skin and then the skin becomes wet from sweat or water, a burn can occur. Many of the burns are slow to notice and by the time a worker feels the burn, the skin is already badly damaged.

Another disadvantage to caustic pastes is the resulting caustic paste "slag" removed from the substrate that must be properly disposed, as it may contain high levels of lead and have corrosive properties. These potentially hazardous characteristics usually render the waste hazardous (see also toxic characteristic leaching procedure, or TCLP testing).

The use of caustic pastes may not apply to all surfaces for reasons beyond those mentioned previously. The application of the pastes may be labor intensive depending on the substrate and conditions affecting the properties of the pastes. For example, temperature and humidity variations may affect some products. For large area applications, some products can be spray applied, thus reducing the labor costs to apply the pastes.

Some manufacturers produce caustic paste varieties containing methylene chloride, a known carcinogen. For this reason, some states may not allow the use of these variations in construction activities, including lead abatement. Currently, the Department of Housing and Urban Development *Interim Guidelines*[1] prohibit the use of on-site methylene chloride stripper. Methylene chloride can also cause carbon monoxide poisoning.

Off-site Chemical Paint Removal Another process used in the removal of lead paint from components is off-site stripping. This process usually occurs at a licensed facility commonly referred to as "dip tank facilities." The paint removal process involves first removing the component from the building and transporting the material to the dip tank facility. Next the components are immersed in vats of chemical strippers, commonly containing methylene chloride. Once the paint on the surface of the components becomes pliable, it is removed and the component surface scrubbed clean of the paint debris and residue. The paint removal process occurs in a controlled work environment with proper worker protection, ventilation, and separation of the multiple-step process.

This procedure is commonly used in stripping the paint from decorative trim to be reinstalled at the project site. Related work includes finish carpentry, prime painting, and finish painting.

Cautionary notes: This process is highly regulated and there are federal Environmental Protection Agency (EPA) licensing requirements. This is because the dip tank facilities are considered a treatment facility by Resource Conservation and Recovery Act (RCRA) definition and because of the use of methylene chloride-containing strippers. Also, this method is more costly than on-site paint removal but is highly efficient in removing all the paint from the component, which is especially useful when deleading properties involved in historical preservation.

Mechanical Paint Removal

New to the lead abatement industry is the use of mechanical paint removal equipment. While this equipment has been experimented with in the industrial painting field (coatings removal from bridges, water tanks, and industrial structures), its application to residential lead abatement has been sporadic. Examples of this equipment are listed below along with a description of their uses.

Needle-guns: This tool is the most common mechanical stripper in residential abatement. It employs the use of numerous long steel rods that impact the painted surface and shatter the paint. It is usually powered by an air compressor and pneumatic hoses. As with most of the mechanical strippers, the needle gun must be used on rigid surfaces such as steel stair components or structural steel components. Noise control is usually required and high vibration of the substrate with adjacent substrates requires added protective measures.

Vibrating, belt and rotary sanders: These types of tools utilize sanding heads to remove the paint from the surface. Their use on steel surfaces is limited but they are effective on wood, concrete, and plaster. However, damage can occur to the substrate if the equipment is improperly used. This equipment can also generate large quantities of lead-containing dust that must be controlled.

Other impact strippers: Various equipment exists that employs the use of steel studs of different sizes and shapes that rotate in an enclosed head to impact the painted surface. As with the needle gun, the paint shatters from the impact. Also similar are the limitations on the use of this type of equipment.

Abrasive blasting equipment: While this equipment is heavily used in the steel coatings removal industry, its use in residential lead abatement is limited. The abrasive can vary from sandblasting, wet sandblasting, soda blasting (bicarbonate), dry ice, and water jet blasting. The process—which uses steel shot, manufactured abrasives, and water blasting—can employ recycling technology that actually separates the paint debris from the abrasive or water for its reuse. Most abrasive blasting is meant for "open air" use, possibly within a contained work area. Innovation in this field has given rise to the development of self-contained abrasive systems that cover a smaller surface area, recycle the abrasive, and are

HEPA equipped. Abrasive blasting of all types can be very dangerous and has limited uses in residential abatement.

The use of mechanically powered abatement equipment in the lead abatement field has required the incorporation of HEPA-equipped vacuum attachments or "shroud." The reason is to provide removal of the respirable dust generated during the use of the equipment. (The term HEPA refers to the filter type on the vacuum used to create the suction in the equipment. A properly maintained HEPA filter will remove 99.997% of all dust, 0.3 microns or greater in size from the vacuum exhaust.)

While the addition of the HEPA shroud has allowed for this type of equipment to be used in lead abatement projects, their limitations vary based upon proper use and application. Most equipment is still dust generating and must be controlled. Even if respirable dust is removed at the source of lead paint removal, the larger dust and chips generated from the paint removal may not be vacuumed into the shroud and will fall to the floor or to surfaces below. Also, many types of tools are equipped with a movable shroud that can be pulled back from the tool for stripping paint in corners or tight areas. Workers have been known to pull back the shroud to view the surface being stripped and to remove the suction that may slow the movement of the tool.

In all cases work area preparation and proper supervision must be employed in the use of any mechanical paint removal equipment. Proper cleaning of all dust generated must occur and varying levels of worker protection will be required. This type of abatement is very labor intensive and the equipment costs are higher than any other method. With the exception of abrasive blasting that is not recycled, the waste generated is relatively low since it usually consists of the paint chips and dust collected in the process.

Manual Paint Removal—Dry Scraping

This method was probably the most common method of paint removal until the effects of lead poisoning caused by improper abatement became evident. A typical part of general construction project is the preparation of surfaces to be repainted. Awareness of lead hazards has increased in painters and individuals in the general construction industry, primarily due to the publicity of lead poisoning in the early 1990s and the recent creation of the Occupational Safety and Health Administration's (OSHA) Lead Standard at 29 *CFR* 1926.62.

Scraping to remove the loose and peeling paint in an effort to control or eliminate the immediate hazard is effective. This temporary solution is commonly recommended as a result of the risk assessment process discussed in the previous chapter.

Manual scraping has several distinct disadvantages:

1. Scraping generates lead paint chips and dust that must be controlled to avoid tracking into adjacent areas.

2. During complete paint removal from a component, the surface is damaged and requires repair.
3. Sanding or "feathering" of paint edges is usually performed as part of this abatement process to eliminate paint lines. This sanding can generate high levels of lead-containing dust.
4. Once a wooden surface has been scraped free of the paint, the next procedure is to sand the wood to remove any marks from the scraping process. However, many wood components such as window sashes and trim have absorbed some of the paint from the initial application. If this has occurred, the final sanding procedure will release lead-containing dust in the sawdust. Unless the sanding procedure is included as an abatement process, with the proper abatement controls, the living space will become recontaminated.

Enclosure

There are numerous conditions where enclosure of the lead painted material is the best option available to remove the hazard from the living environment. In many cases it offers a permanent separation of the hazard from the living space, and in other situations, the enclosure is temporary at best. It is important to make the distinction between terms used in the abatement industry that may promise a long-term solution such as "encapsulate" and "permanent." An enclosure must be instituted as an extension of the building structure, exhibiting the permanence of the structure itself. A "permanent" enclosure should last the life of the building or at least the life of the function of the building component it is enclosing. For example, if the exterior trim of windows are enclosed within the shell of the installed, prefabricated window, the trim is essentially enclosed for the life of the window/building structure. Also, if gypsum wallboard is installed to enclose a lead-painted wall, the enclosure is effectively permanent, at least until the plane of the wall is penetrated in the future.

The permanence of an enclosure must be considered in all cases of recommending this abatement procedure. Each condition of enclosure will have varying limitations, as exhibited by the examples above.

The life of an enclosure may also be increased by periodic monitoring of its condition. For example, covering an exterior window sill with heavy gauge aluminum will serve to render its lead paint inaccessible. However, vandalism or damage to the aluminum will shorten the lifespan of the enclosure.

While enclosing lead-based paint may be extremely cost-effective for many situations, the hazard has only been rendered inaccessible for a finite period. Long-term costs must be considered if future renovations affect the enclosed surface or component. Examples of enclosures include:

- Gypsum wallboard covering a wall
- Metal panning, prefabricated and sealed to the surface such as window sills, casing, and trim

- Re-siding over existing exterior wood siding
- Vinyl siding of exterior wood siding (best effective if the existing paint is rendered intact and the siding is insulated with a weather protectant material between the original siding and the vinyl)
- Prefabricated door casing installed over the existing opening

Encapsulate

To encapsulate, by definition, is to enclose. However, in the lead abatement industry, as in the asbestos abatement industry, the term has been broadly used to include products that could be classified as binding products, paints, and coverings that may not offer enclosure properties for the long term. Encapsulant products must be tested and guarantees offered by the manufacturer.

Encapsulants should be judged on their binding capabilities to render the lead-based paint inaccessible. Impact damage, vandalism, and normal wear over time must be considered and tested for each product. Product demonstrations are highly recommended and test areas should be applied and monitored to determine if an encapsulant will stand up to its surrounding environment. Many encapsulants that bind to the surface have long-term curing times, sometimes through the seasonal change of the area. Some can become as effective as replastering a wall or coating it with impenetrable vinyl or rubber. Other encapsulants are similar to reinforced wallpaper or paint and will fail in a very short period of time. For this reason, the effectiveness of an encapsulant is greatly affected by the condition of the paint and the surface. For example, many encapsulants will fail if sealed to a damaged surface such as nonintact paint or damaged plaster.

Encapsulants should be considered a short-term solution to render the lead paint inaccessible. The life of the encapsulant will be determined by the accessibility and condition of the surface covered. Moving components or friction components on windows and doors should not be encapsulated. Wall corners, counter tops, stair components, and floors are generally high-traffic and high-damage areas and are not recommended for encapsulation.

Comparison of Methods

As stated previously, the abatement methods discussed represent differing approaches to the primary goal of removing the hazardous condition in the residence. Some methods provide for the complete removal of the lead-based paint such as component removal and paint removal methods. Other methods, such as enclosure, attempt to render the component with lead-based paint completely inaccessible within the building structure itself. Finally, encapsulants offer a short-term method of covering the lead-based paint with products that adhere to the painted surface and render the surface inaccessible.

As discussed, the planner must make a comparison of methods by reviewing the applicability for the substrate(s) to be abated, as well as the long- and short-term costs and savings of each available method.

ABATEMENT PROCEDURES

Lead-based paint abatement is comprised of four basic elements: work area preparation/isolation, the appropriate abatement method, cleaning activities, and final visual inspection and clearance. Acceptable work procedures have been developed from basic industrial hygiene principals of proper administrative controls, engineering controls, and worker protection. For example, isolation of the work areas, decontamination facilities, and worker training in the hazards of lead abatement are some administrative controls of lead abatement. Engineering controls such as HEPA-equipped vacuums, abatement equipment with HEPA shrouds, and state-of-the-art cleaning equipment have been incorporated into the lead abatement industry from others such as the asbestos abatement industry and the industrial coatings removal industry. Worker protection such as the various levels of respiratory protection and protective clothing have also been incorporated into the lead abatement industry from related industries.

The procedures involved in proper lead abatement are discussed with the intent that administrative and engineering controls shall be utilized to reduce the amount of worker protection needed for the performance of the abatement. It is the responsibility of the planner to design and recommend state-of-the-art procedures and abatement methods and to require the use of procedures that have become standards in the lead abatement industry. It is the role of the abatement contractor to use the methods available in such a manner as to limit or reduce worker exposure to lead and to remove the lead hazard as the completion of the abatement.

The basic steps of any lead abatement project start with work area preparation and isolation. This step is required to segregate the work area from adjacent areas for many reasons depending upon the abatement situation. For example, isolation is required to ensure cross-contamination of adjacent spaces does not occur. However, it also provides for a well-defined separation of trades, as is desirable for control on a large scale project. Lead abatement methods generally create lead-containing dusts that can travel with air currents or be tracked by workers passing through an abatement work zone. Also, designated work areas are effective to ensure the completion of abatement prior to other trades entering the work area and performing other tasks. Thus abatement is treated like a separate phase of work that must be inspected and approved prior to future work.

Work area preparation includes the following:

- Preclean to remove all dust and debris by using an HEPA-equipped vacuum. In many cases, lead-containing dusts exist on the surfaces of the abatement work area prior to any abatement being performed. This may be from dust generated from friction surfaces such as opening and closing window sashes or doors. The lead may exist in exterior soil that is continually tracked into the house or it may be airborne in dusts from a variety of sources exterior to the home. A lead abatement contractor should preclean the work area to remove any loose paint chips, dust, or debris from the floors, window sills, and other horizontal surfaces. Only then should the next steps of work area preparation be performed. Any dust generated in the work area after precleaning activities will be due to the abatement work, thus allowing for a baseline to measure from in monitoring the abatement work.

- Cover floors of the work area with two layers of polyethylene sheeting. This will contain lead dust for clean-up after the abatement process. Also, lead-containing dusts can get into cracks in the floor only to be released back into the living environment later.
- Seal all heating, ventilation and air conditioning (HVAC) units and associated grilles and vents airtight. This will prevent cross-contamination into adjacent spaces. Power to the HVAC serving the abatement area should also be shut down.
- Seal all access to the work area including windows, doors to adjacent rooms and to the exterior (termed critical barriers). This will prevent cross contamination but will also stop or slow air movement in the work area, allowing generated dust to settle.
- At the entrance to the work area from the exterior, construct a decontamination chamber. This should consist of a minimum of two rooms, a clean room (closest to the exterior) and an equipment room (closest to the work area). The equipment room is used for removal of dirty clothes and cleaning and bagging of equipment. The clean room is designated for workers to wash and change into clean noncontaminated clothing. While a three-chamber decontamination chamber is not currently incorporated into most existing regulations and guidelines for lead abatement, it is recommended. The three chambers consist of the same two as stated previously with a shower placed between for proper decontamination. Basic hygiene such as washing all exposed areas with soap and warm water will eliminate exposure to lead-containing dusts from hand-to-mouth activities such as eating or smoking. It is for this reason that the three-chamber decontamination chamber is recommended. Supervision of the workers to ensure proper decontamination will serve to keep lead-containing dusts inside the work area.
- Post warning signs at the entrance to the work area and at all accessible critical barriers. This will warn others that a secured work zone is in place and only authorized personnel are allowed to enter.
- The HUD *Interim Guidelines*[2] require the use of ventilation in every abatement work area. However, this ventilation is not required to be filtered. In many abatement situations involving some types of component removal, enclosure or

encapsulation, this may be adequate. However, if the method of abatement involves paint removal, dust-generating demolition, or scraping of loose lead paint, it is recommended that HEPA filter ventilation be utilized. This will eliminate the dust generated from being ventilated to the exterior of the work area, thus contaminating outside spaces. In cases of higher levels of dust generation such as the use of power tools or abrasive blasting, it is strongly recommended that the HEPA-equipped filtration units be used to create a pressure differential between the work area and the surrounding spaces.

PERFORMING THE ABATEMENT

The contractor performing the abatement should be trained in all aspects of lead abatement including proper worker protection, work area isolation, abatement methods, and cleaning. The HUD guidelines describe in detail, the requirements of a three- to four-day training course required for all contractors performing lead abatement work on federal projects. This training requirement has been adopted and enhanced by many states formulating regulations such as Connecticut and Rhode Island. Workers are also required to receive medical and blood monitoring prior to performing lead abatement and periodically thereafter.

The experience of an abatement contractor is crucial to a successful abatement process. The proper work area containment, cleaning procedures, and clearance criterion requires training in the proper use of equipment and cleaning methods discussed in this section.

Proper supervision of abatement workers is also very important. If any step in the abatement process is missed, the result can be a work area that does not pass the clearance criteria, requiring repeated cleaning. For example, if precleaning of the work area and adjacent affected areas is not performed, further tracking of lead containing dusts can occur, increasing the scope of cleaning needed. Proper use of equipment and proper abatement procedures must be employed or uncontrollable amounts of lead containing dusts may be generated. For example, if a worker attempts to expedite the paint removal procedures by using a belt sander to remove lead paint, this will result in the generation of large quantities of lead containing dust. The work area may not be equipped for this type of abatement procedure and this dust may contaminate other areas.

Some basic procedures in lead abatement are as follows:

1. Remove all movable items out of the room to be abated. Carpeting, drapes, and/or wall fixtures can be removed and cleaned.
2. Tenants should not occupy the premises until the abatement is completed and the clearance criteria are met.
3. Other trades should not be allowed to enter the work area at any time during any phase of the abatement.

4. Workers should not smoke, eat, or drink inside the work area.
5. Daily cleaning procedures as described here should be standard procedures for every job, no matter the size.
6. Decontamination procedures must be followed to reduce worker exposure and eliminate tracking of lead contamination into adjacent areas.
7. All debris, garbage, and worker coveralls should be bagged daily and removed from the site.

Worker protection during the abatement operation must be based on exposure levels to the workers for the task at hand. As described in this text, worker exposure is tested prior to reducing the level of respiratory protection. Once exposure levels are established and are reproducible, work may commence for a given procedure.

At the end of every work shift, the abatement crew should stop abatement activities to leave enough time for general cleaning of the work area. Many lead abatement regulations require the removal of all generated debris from the work area daily. This debris may consist of demolition debris, paint chips, or collected dust. All of this material should be placed in 6-mil polyethylene bags, sealed and removed from the work site daily. This prevents a build-up of debris that is tracked out of the work area, decreasing the chance of cross-contamination.

FINAL CLEANING PROCEDURES

The final step of the abatement procedure is actually composed of several steps. Once removal of the bulk debris is completed, the abatement contractor should mist down the work area to accelerate the settling of the dust, then remove the floor polyethylene sheeting by folding the sheeting into itself, containing any remaining debris. The contractor shall then leave the work area for 24 hours to allow any airborne dust to settle.

Upon returning, the contractor will then perform what is commonly termed the HEPA-TSP-HEPA cleaning. This is a three-step procedure that begins with a thorough HEPA vacuuming of all surfaces of the work area. The next step is a complete washdown of the work area surfaces with a solution of tri-sodium phosphate (TSP) soap and water. The solution is 1 ounce of 5 percent TSP solution in 1 gallon of water. The final step is a repeated HEPA vacuuming of all surfaces and wipedown of the surfaces to remove any residue present.

In this final cleaning phase of the project, it is very important to wait the 24 hours and then perform all three steps of the procedure. Airborne dust will take overnight to settle if there are operations such as sanding involved in the abatement process. Misting the air with water atomizing devices will obviously expedite settling. However important this last step of cleaning is, it is usually the first corner cut by a contractor who is attempting to save costs. The result is that the contractor has only created more work (and cost) by having to repeat the cleaning if the clearance criteria are not met.

FINAL INSPECTION AND CLEARANCE CRITERIA

At the completion of the final cleaning procedure, an inspection of the abatement work area should be conducted prior to release of the work area to the owner or other trades. This inspection is required by many state regulations and by HUD for federal public housing projects. An additional requirement is that the inspector be a third party to the owner/contractor agreement, hired by the owner.

The final inspection consists of a review of the work performed, a visual assessment of the cleanliness of the work area, and dust testing. Also termed the abatement reinspection, the following should be conducted as part of the final inspection:

- Refer to the original lead testing report of specific scope of work that identifies where lead-based paint existed in the work area. Use this document to inspect the work area to ensure all of the identified lead-based paint has been abated. Refer also to the scope of work documents to inspect for the correct abatement procedure. Other work items related to the abatement must be checked in this process. For example, the proper use of an encapsulant product or the proper installation of the enclosure material should be inspected as part of the final inspection. New installed materials should not be painted prior to the final inspection. This visual inspection may occur prior to the 24-hour settling period to inspect for work completion and proper abatement procedures

- Inspect the work area for visible dust or debris. The HEPA-TSP-HEPA procedure should remove all visible dust from the work area. Any signs of visible dust should be directed to the contractor's attention for recleaning. This inspection must occur after the 24-hour settling period.

- Next, perform dust testing in the work area. The formal protocol for dust testing by HUD should be referenced. This testing measures the lead in dust after abatement and compares these results to clearance criteria set forth in the contract documents. Specifically, HUD and many states have adopted the following clearance criteria:

  200 micrograms lead per square foot for floors
  500 micrograms per square foot for window sills
  800 micrograms per square foot for window troughs (wells)

These clearance levels represent the acceptable amount of lead in a square foot area of tested surface. It should be noted that when proper cleaning is performed, levels of 15–30 micrograms lead per square foot to as low as none detected are achievable.

Dust testing should be performed, as a minimum, in every room where abatement was performed. In some rooms, such as kitchens and children's rooms, testing may be repeated on additional surfaces such as countertops, shelves, window sills, and so on. The dust tests (commonly termed wipe tests) are analyzed by a laboratory for the amount of lead present on the dust wipe medium. Wipe tests are commonly performed using alcohol-free baby wipes purchased at any grocery store.

The inspector wears sanitary gloves for each test to eliminate cross-contamination and wipes the surface in a measured area, usually 1 square foot. The wipe is then placed in a sealed, sanitary container and delivered to the laboratory for analysis.

There are available wipe testing kits and dust testing kits at hardware stores and other places. While these test kits may be good indicators of the presence of lead, their reliability and relationship to the established clearance criteria is yet to be known.

Once a work area has been accepted and cleared for reoccupancy, proper record keeping must follow. The owner should have a record of the abatement by retaining the initial inspection report, the abatement plan and updated as-built drawings. As-built drawings should record location of lead-based paint remaining, encapsulated or enclosed. New construction materials should also be recorded, especially if partial abatement was performed. The abatement contractor should also submit to the owner the receipts and warranties for the replacement materials (window and door systems), warranty for the contractor's work, and the receipts and manifests for the waste disposal related to the project.

WASTE DISPOSAL

The waste generated from lead-based paint abatement must be disposed of properly. The disposal must follow local, state, and federal regulation regarding disposal of solid waste. The current interpretation of the federal Resource Conservation Recovery Act (RCRA) Solid Waste Regulations requires the waste generated to be tested for the toxicity characteristic of the lead in the lead-based paint.[3] The required test is the Toxicity Characteristic Leaching Procedure (TCLP), which measures the leachable quantity of lead in the components to be disposed. With lead known to exist in the waste material and with lead a known hazardous material, the waste is subject to TCLP testing under the RCRA regulations. The TCLP testing will determine whether the waste must be classified as hazardous because of its lead content or if it can be disposed in a construction landfill. The regulatory limit for lead toxicity according to the TCLP test is 5.0 mg/L.

The cost for TCLP analysis, per sample submitted, ranges from $75 to $150, depending on turn-around time requested from the laboratory. A full screen TCLP is often requested by the site accepting the hazardous waste. This test includes the toxicity testing for all eight RCRA metals (including lead) as well as testing for flammability, corrosivity, and pH. The cost for the full screen TCLP test ranges from $350 to $600 per sample submitted.

The TCLP test is performed by retrieving a representative sample of the waste to be disposed and submitting it for laboratory analysis. The laboratory then reduces the size of the sample to fit through a 9 millimeter sieve and immerses the material in an acidic solution of known pH. The solution is then agitated for 15 hours, then analyzed for lead content, with the lab measuring the amount of leachable lead in solution from the waste material.

The TCLP test must be performed carefully on a lead abatement project since the cost of hazardous waste disposal is much greater than nonhazardous waste. For example, a 40 cubic yard dumpster full of door trim, windows, and window trim as hazardous waste would cost approximately $12,000 to $14,000 to properly dispose at an EPA-licensed hazardous waste treatment and disposal facility. That same amount of material, if nonhazardous, costs $500 to $1,000 to dispose as regular construction debris. This obviously represents a great difference in cost, but there are measures to reduce these costs and limit the amount of hazardous waste generated at a site. These methods are segregation of waste and waste reduction techniques.

Segregation of waste refers to procedures to identify those components that fail the TCLP test criteria and separate those materials from the waste stream to reduce the overall volume of hazardous waste. For example, if waste materials in a dumpster of abated windows and doors are tested separately and it is found that the window sashes and window trim fail the TCLP test but the doors and door trim pass when tested separately, these materials should be separated for disposal. Also, the lead inspection testing results do not correlate to the TCLP test and cannot be used to determine lead leachability.

For waste segregation, TCLP testing should be performed prior to demolition procedures to avoid expending labor to separate waste from dumpsters already filled with abatement wastes.

Waste reduction is a relatively new technique to reduce the overall volume of classified hazardous waste. These procedures are not to be confused with waste treatment, which renders the waste nonhazardous. Waste treatment can only be performed by an EPA-licensed treatment facility. Waste reduction procedures reduce the volume of waste by the following methods:

- Processing the waste material through a modified wood chipper enclosed in a sealed work area.
- Remove the paint from the substrate with the use of a wood lathe or abrasive blasting inside a modified, sealed work area.

- Breaking up the waste into smaller, more compact pieces to be packed into the dumpster, thus fitting more waste into the dumpster for the cost of the dumpster. A modified compacting of the waste in the dumpster is also performed.

All paint debris generated from the reduction process must still be disposed as hazardous waste since treatment is not performed on the hazardous constituent, the lead-based paint. These techniques can dramatically reduce the cost of disposal and it is recommended they be employed where feasible.

RELATED WORK AND ARCHITECT'S ROLE

The role of an architect in lead abatement projects is very important in large-scale projects. The architect must design the integration of the lead abatement into the other work involved in the project, which at times differs greatly from previously performed construction methods. The installation of window or door systems after abatement has occurred usually must be performed the same day as their removal for security reasons. Cleaning and clearance of the work area must be performed prior to the installation of the new materials.

The installation of drywall enclosures or facades over lead-painted material is usually specified by a project architect but consideration must be given to the long-term lead abatement goals of the project. Painting procedures can no longer include scraping of the lead paint by the painters as part of "surface preparation."

The architect should also make all contractors aware that lead-based paint exists in the scope of the project and the general contractor and deleading subcontractor must be held responsible for this coordination. This information is usually included in the "General Conditions" section of the project specifications and warning language included in all other sections. The more the contractors are aware of the presence of lead-based paint, the better coordination between trades.

The final outcome of an abatement project should be a living environment free of hazardous lead containing dusts and accessible lead-based paint. This is best achieved through coordination of the design activities with the lead consultant and the architect, making all trades aware of the abatement process as part of the project schedule. As is similar with a general contractor overseeing the procedures of the subcontractors, the project architect is usually responsible for coordinating all subconsultants, including the lead consultant.

The architect may also be coordinating project phasing with the owner and an owner's representative for tenant relocation, or "relocation specialist." This person has a very important role in coordinating the relocation procedures with the tenants, tenant task forces, and community groups. For tenants, the relocation specialist is the primary contact with the owner and architect and also serves as a communication channel for the architect to the tenants, thus making the relocation procedures run smoothly.

Relocating tenants temporarily during the abatement process is painful to all involved if the following procedures are not followed:

- Inform the tenants during the design process. This means involving tenants' ideas in the design of the project including materials, color schemes, and coordination of moving tenants' belongings, wherever possible.
- Inform all tenants of the contractor selected to perform the work and the work schedule. Inform them of the importance of their role in cooperating with the relocation specialist and movers.
- Inform them of the lead abatement procedures and the result of the abatement process. Some of this can be conveyed through a tenant meeting with the assistance of the architect and consultant.
- During the process, the relocation specialist will be the primary contact, fielding all concerns of the tenants and owner. Periodic meetings should be held to review the procedures and any problems in relocation.

The architect should require an experienced lead abatement contractor to perform the work of the project. Obtaining an experienced contractor is possible through prequalifying procedures based on previous work experience and successful completion of projects similar in scope and magnitude to the project at hand. The architect can require the general contractor to select only an abatement contractor with previous work experience in lead abatement. This can be a required submittal prior to award of the contract or approval to start the work.

Review of project submittals, as in any construction project, are an important step in approval to start the work. The architect is responsible to review the submittals and approve them quickly to allow the contractors to order materials that have long lead times. Prefabricated windows and door systems can require up to eight weeks for delivery on site. In many projects, abatement cannot occur until replacement materials are delivered to the site. If incorrect materials are approved, ordered, or installed, the abatement process, relocation efforts, and the project schedule can be greatly affected.

Prior to the start of large-scale abatement projects, the architect and lead consultant must perform intensive field investigations. This must be done during the design phase, since the results may affect the schedule and the order of work performed. The field investigation must include existing conditions of the building materials and how this will affect the work. For example, if severe damage has occurred to the dwellings due to fire, vandalism, or deterioration of building materials, the scope of work may change. The field investigation should also include discussion with the site maintenance personnel who know the hidden secrets of the site, (termite problems, leaking roofs or windows, locations of fire damage or vandalism).

If these conditions are considered during the design phases and worked into the project, the possibility of change orders due to unknown field conditions is less likely.

Notes

1. U.S. Department of Housing and Urban Development. *Lead-Based Paint: Interim Guidelines for Hazard Identification and Abatement in Public and Indian Housing.* Washington, DC, September 1990.
2. Ibid.
3. Toxicity characteristic final rule, *Fed. Reg.* March 29, 1990 (55 *FR* 11798).

11

Abatement Case Studies

Keith W. Pokorny

The following case studies are taken from actual abatement projects occurring at large-scale housing authorities and in smaller residential projects. Suggested solutions and problem solving are offered to assist project planners, owners, administrators, architects, and consultants in the proper design and planning of an abatement project. Some of these options are offered in hindsight, and it is hoped that others can learn from the experiences and problems encountered in previous lead abatement projects.

BIDDING HAZARDOUS WASTE DISPOSAL

It is most cost-effective for the owner to obtain bids for hazardous waste disposal if that cost is included in the base bid for the entire project. An example of this is two public housing authority (PHA) projects that bid this work two different ways. The first PHA informed the contractor in the project specifications that the cost for hazardous waste disposal would be included as a change order to the project once TCLP testing is performed and determined during abatement. The PHA retained the right to seek alternative prices if the change order proposal was not acceptable.

The second PHA decided to identify the hazardous waste prior to bidding the project. The language of the specification required waste reduction procedures to minimize the volume of waste. The decision by PHA 2 to include the waste disposal costs for all waste reduction, permit costs, hauling of the waste, and all other costs resulted in requiring the contractors to be competitive in the waste costs, provide waste reduction, and account for all fees involved in the process. For PHA 1, it incurred the general contractor's 10 percent mark-up on the change order, and waste reduction costs could not be realized since the procedure was relatively new.

Another method to bid the waste reduction is to also request unit prices for the disposal of a 40-cubic-yard dumpster of waste. This would serve as a set price to be used in additional work and credits for this work.

APPROVAL OF PRODUCT SUBMITTALS BEFORE ABATEMENT

As discussed in Chapter 10, the architect must be very careful to review the submitted product submittals prior to allowing work to commence. For example, a PHA required the installation of window materials on the same day as the leaded windows were removed and cleaning was performed. An on-site inspector was present to review the window opening and inspect for visible dust. However, the architect for the project had allowed the contractor to be delinquent in the product submittals for the windows to be installed.

The result was the contractor rushing to get approval of the windows for the start of this phase of the project. The windows submitted were incorrect for the resulting opening and were rejected by the architect. Unfortunately, other work on the project had begun. The contractor had to order different windows, incurring a four- to six-week delay. The delay affected the order of work and the project schedule. This extended the project time into the winter months because abatement could not occur on the windows until the new materials were approved, ordered, and on-site.

USE OF CAUSTIC PASTE ON WOOD COMPONENTS

Lead-based paint existed on the interior wood casings of a home. The abatement contractor used caustic paste to remove the paint since the window trim was decorative and the owner wanted to retain the trim in good condition. The work area set-up was correct, as was the application of the caustic paste. The contractor removed the paste effectively and neutralized the wood surface to complete the procedure. However, before the wood dried completely, the contractor removed the work area containment around the windows that was constructed to isolate the work from the room.

When the wood dried, there was a visible crystalline film on its surface. Unknown to the contractor, this material was the result of the caustic paste and neutralization solution seeping into the wood and then drying out, producing the film. The film contained lead due to the caustic's ability to chemically break down the paint. The contractor proceeded to remove the film by sanding the surface with finish sand paper. Since the work area containment was already removed, this resulted in contamination of the rooms, requiring additional cleaning by the contractor.

WHEN CLEARANCE TESTING FAILS TO MEET CRITERIA

Lead abatement was performed in a day care center in several locations. One location was the removal of the basement windows from the lower level rooms. This procedure consisted of the removal of all of the window components, leaving behind the concrete opening for a new window to be installed. Once the window sill was removed the window stool was removed, which exposed the rough concrete opening. The surface was level enough to allow for the installation of a new window; however, the surface was rough with cracks and grooves that had trapped and collected dust from the window over the years of operation.

As a final step in the deleading process, the contractor performed HEPA-TSP-HEPA (high efficiency particle air–tri-sodium phosphate) cleaning of this grooved surface to prepare it for a final inspection and wipe testing by the inspector. Wipe testing showed the samples failed the clearance criteria by exceeding the specifications predesignated value of 200 micrograms per square foot (the actual results exceeded 1,000 mg/ft^2). Although cleaning was repeated and wipe testing performed three times, the level never reached below 200 μg/ft^2. It was discovered that the grooved surface had trapped the dust and debris so that with each cleaning more dust rose to the surface and became accessible to the wipe test.

A solution to resolve this was for the contractor to perform HEPA-TSP-HEPA cleaning for each window opening and the inspector to perform a visual inspection of the concrete sill. Then the contractor covered the surface with a urethane coating to seal it prior to the installation of the new window blocking. Wipe testing was performed on the urethane coating of the next three windows installed and all passed the clearance level of 200 μg/ft^2.

CONTRACTOR'S RESPONSIBILITIES

In the specifications, if the contractor does not meet the clearance criteria after the first cleaning, designate the contractor responsible for the cost of the additional sampling and analysis. While not all projects will pass the clearance criteria after the first cleaning, most should if the cleaning is performed properly. Grooved surfaces such as concrete and masonry are more difficult to clean thoroughly, as described above. However, an owner should not incur the cost of poor cleaning by the contractor. Also, having the contractor responsible for the additional testing costs may provide incentive to the contractor to properly clean the work area. It will become evident when the contractor fails to use TSP or tries to cut corners in the cleaning process.

KNOWLEDGE OF PREEXISTING LEAD LEVELS IN DUST

As part of any abatement project, an owner and contractor should be aware of any preexisting levels of lead in the dust. An example of this affecting the work of a modernization project occurred at a housing authority development where there were high levels of lead on the basement floors. The levels were found to be in excess of 10,000 micrograms lead per square foot. At this development, all of the abatement was performed in the upper floors of the three-story buildings. However, all of the replacement materials—such as windows, trim materials, flooring tiles, and bathtubs—were stored in these basements upon delivery and prior to their installation. As a result of discovering high levels of lead dust after the new materials were installed, a thorough final cleaning was instituted to remove any dust tracked into the development from the basement areas. This cleaning was referred to as the reoccupancy cleaning, and it occurred after installation of the new materials and prior to finish painting of the units.

REVIEW OF THE INITIAL LEAD TESTING

Many lead inspectors are not properly trained to understand the results their instrument is giving them, therefore a thorough review of the initial testing must be done. One example of this is the concept of "read-through"—when the X-ray fluorescence (XRF) result indicates a positive reading on a negative surface due to the existence of lead based paint on the underlying surface on which the tested component is attached. For example, if a wall of a dwelling contains lead paint at an XRF level of 10.0 mg/cm^2, the result of testing the chair rail over the wall may reveal a reading of 2.0 mg/cm^2. Also, window trim such as casings or aprons placed over this same painted wall may exhibit similar readings. This can result in the abatement of components that test negative for lead paint content.

A solution is to make sure that when overlapping components exist on a leaded component, the XRF testing includes the bare substrate of the top component. This is commonly referred to as the substrate equivalent lead (SEL) reading. Chemical spot testing with 6–8 percent sodium sulfide solution is also a good check for "false positives" due to read-through.

IMPROPER USE OF ENCAPSULANT PRODUCT ON FRICTION SURFACES

One property owner relied on the promises of the manufacturer of an encapsulant product in making its decision on the appropriate abatement methods for windows and doors in the development. The owner was convinced that an encapsulant product (a fiber cloth mesh with an epoxy resin) would be suitable to encapsulate the lead paint on the interior door and window trim components. Unfortunately, within weeks of the installation of the product, the opening and closing of the windows and doors impacted the seal of the product. Also, the installation of the product was poor and delamination occurred due to temperature and humidity changes over the season.

The owner had installed the product in the entire development. The owner later chose to enclose the entire window casing and door jamb with a prefabricated aluminum covering sealed and face-nailed to the substrate.

This example shows the improper application of an encapsulant product that may actually be effective if installed properly and over flat surfaces, such as walls or ceilings. The use of any encapsulant or enclosure on movable parts such as doors or windows is not recommended unless prior testing is performed to prove its durability. For a similar cost, the paint can be removed for a long-term solution.

PREVIOUS ABATEMENT OR RENOVATION

In many housing developments, especially public housing that has undergone comprehensive modernization or upgrade, previous work may have included abating lead hazards. It is essential to review the type of work previously performed before undertaking new work because the previous abatement may not have been performed to current standards. Also, if prior modernization did not include considerations for lead paint, the new work may uncover an old hazard.

A public housing authority in the Northeast had purchased a building to renovate into public housing. Previous renovations had occurred to the exterior, including the application of vinyl siding over the existing clapboard siding. As part of the planned renovations upon purchasing the property, the housing authority was to remove the vinyl siding where damaged and replace it with new siding to match. As a part of this procedure, the contractor would be requested to remove the existing weather proofing under the vinyl siding and replace it. This weather proofing was attached over the lead-painted clapboard, which was discovered to be chipping and peeling. Apparently, the contractor who had installed the siding had failed to remove the loose and peeling paint and render the painted surface intact.

NEW PRODUCT PROMISES

In a project specification, a lead abatement contractor was allowed to present alternative options to the owner on the most cost-effective method to remove lead-based paint from plaster wall surfaces. The contractor chose three methods: caustic paste application, a rotary sander with HEPA attachment, and a vibrating sander with HEPA attachment. The contractor presented each method in separated rooms of the building on the same floor. Each room was monitored for dust generation, both visually and with worker exposure testing.

The two HEPA equipped sanding tools never made it past the first day of monitoring for several reasons. First, the dust generation was much greater than the manufacturer's promise of the HEPA vacuum attachment. Visible dust coated the room's floor. Secondly, the sanding pads for the vibrating sander could not penetrate the lead paint surface. The rotary sander could, but it took great pressure from the worker. When each of the sanders did penetrate and remove the lead paint, the pressure was so great on the sander that the worker kept gouging the plaster wall such that the repair needed would be extensive.

The caustic paste removed all of the paint to the substrate and the worker exposure was minimal.

MECHANICAL STRIPPERS

I issue a note of caution about the use of mechanical strippers and their effect on adjacent surfaces. A PHA was approached by its deleader with a proposal for alternating the abatement methods on the steel door jambs at the apartment entry and within the apartments. The current method used was caustic paste. The contractor was proposing the use of the needle-gun with a HEPA attached shroud.

After the proper set-up, preparation of the room for abatement and ear plugs for noise protection, the contractor proceeded to impact the door jamb with the needle-gun. The tool removed the paint sufficiently and prepared the surface for repainting very well. It was not until someone leaned against the door jamb that it became apparent that the impact from the needle-gun had detached the door jamb from its concrete setting. Almost a full quarter of an inch of space now existed at what was the seal of the door jamb to the plaster wall. It was then decided by all present to stay with the caustic paste method of paint removal.

12

Removing Lead-based Paint from Steel Structures

Lloyd Smith, Ph.D.

The major reason for painting steel structures is corrosion protection. Corrosion is an electrochemical reaction. This requires moisture on the steel surface for the corrosion reaction to take place. Coatings act as a barrier to isolate the steel from moisture and other contaminants. But no coating is impermeable. Therefore inhibitive pigments were added to the coating formulation. These pigments either react with the contaminants or migrate to the substrate surface and passivate the steel. Red lead and related lead compounds have proven themselves to be effective inhibitive pigments for prolonging the life of the coating system. A study conducted by Steel Structures Painting Council in the early 1980s showed that paint formulations available at that time with alternative pigments did not perform as well as the lead-based pigment control formulation.[1]

Coating systems based on lead-containing primers have a long, proven history of performance. The highest concentrations of lead in a paint can be found in lead-based primers for protection of steel structures. An example is TT-P-86, a U.S. government formulation specification for ready-mix, red lead-base primer.[2] There are four formulations in this specification, with the minimum amount of pigment varying from 65 percent to 77 percent by weight. The minimum amount of red lead in the pigment varies from 65 percent to 99.6 percent. Total lead concentrations for samples taken from steel structures which were painted with coating systems containing lead-based primers typically range from 10 percent to 50 percent.

EXTENT OF THE PROBLEM

While lead-based primers are rarely used today and have been replaced by other materials, the number of existing steel structures with lead-based primers is enormous. For example, Federal Highway Administration estimates that 89.2 percent of the steel bridges carrying public roads have lead primers on them.[3] This translates to 185,900 of the 208,500 steel bridges in the National Bridge Inventory. CONSAD Research Corporation reported 51,645 steel water tanks are operated by community water systems to hold drinking water.[4] It is estimated that 65 percent of these tanks, or 33,500, were painted with lead-based primers. This does not include industrial water tanks, which would most likely double the water tank estimate. The corporation also estimated that 209,350 of the 700,000 petroleum storage tanks in the United States are coated with lead-based primer. The total number of steel structures protected with lead-based primers has to be much higher, as estimates for the marine industry, chemical plants, power plants, government facilities, and other structural steel are not available. And these estimates are only for structures painted with lead-based primers. Other steel structures will have coatings on them that contain lead, albeit at a lower concentration, in the form of driers or coloring pigments in the paint.

SURFACE PREPARATION

As the primary intent of the coating system on steel structures is corrosion protection, decisions on maintenance or replacement of the coating system are usually based on corrosion or aesthetic considerations and not on lead removal. The term "lead paint abatement" is a misnomer in the context of steel structures. Abatement means a lessening or reduction. Very few steel structures are repainted solely for the purpose of reducing the amount of lead on the surface. Maintenance funds have generally been limited. The added cost to comply with lead issues has resulted in even fewer steel structures being painted each year.

Surface preparation is one of, if not the most important, aspect that determines coating life. A new coating system will provide longer protection on a well-prepared surface compared to a marginally prepared surface. In addition, some of the newer coating systems require that the surface be thoroughly cleaned to perform as intended. Specifications for surface preparation of high levels of cleanliness, such as the Steel Structures Painting Council (SSPC) or National Association of Corrosion Engineers (NACE) Abrasive Blasting Specifications, require all old paint, rust, mill scale, and other contaminants be removed from the surface. All that is allowed to remain are specified quantities of stains or discolorations of old paint, rust, and mill scale.

Mill scale is an oxide of iron formed during the hot rolling process when steel is formed into plates. The mill scale is hard, brittle, and impermeable. It was considered to be a protective coating for many years. Therefore, it was common practice to leave the mill scale on the surface and paint over it with a lead-pigmented, oil-based primer. However, the mill scale cracks as it ages upon exposure to the environment. Moisture can then get in and set up a corrosion cell. Since mill scale is more noble than steel, the steel underneath the mill scale corrodes. Once the rust forms, there is no adhesion between the mill scale and the steel, so the mill scale pops off, taking the coating system with it. This was recognized as a problem, and it became popular to remove the mill scale prior to coating starting in the 1950s and 1960s. In addition, the development of high performance coating systems such as those based on zinc-rich primers require intimate contact between metallic iron and the zinc metal dust in the primer to obtain the long-term performance of these coating systems. Today, it is quite rare to leave the mill scale on the surface of steel that will be exposed outdoors.

Abrasive blasting is the preferred method of surface preparation when high levels of cleanliness are needed. The process not only removes the old coating, rust, mill scale, and other contaminants but also roughens the surface, which promotes adhesion of the new coating system. A distinction must be made between lead paint removal and surface preparation. As the important aspect for corrosion protection is to properly prepare the surface, lead paint removal has been a secondary consideration. Processes are available to remove the lead-containing paint but if a high level of surface preparation is required, a two-step process of lead paint removal followed by surface preparation significantly adds to the cost of the project. The preferred method is to perform both steps in one operation.

Except for the mildest surface preparation methods such as hand-tool cleaning, most other surface preparation methods are quite aggressive. In most other cases, the intent of the surface preparation method is to remove hard materials such as rust and mill scale and to cut into the steel to roughen the surface. Processes that perform this function pulverize a good percentage of the paint, forming a fine dust and thus increasing the potential of worker exposure to high levels of lead.

COATING SYSTEM MAINTENANCE

Due to a lack of understanding in the industrial painting industry of the issues involved regarding lead paint removal, the significant changes in procedures and methods, and cost, there has been a resurgence in coating system maintenance as opposed to coating system removal. Coating system maintenance consists of spot repairs to the coating system using mild surface preparation methods such as hand-tool cleaning, usually with the application of a new top coat to the whole structure. The intent is to prolong the life of the coating system with the hope that technology will develop better and/or more cost-effective methods of removing the paint, or that lead paint removal methods will become more standardized.

Maintaining a coating system is a decision that must be made carefully. Coatings degrade with time and do reach a point where their adhesion or integrity is such that maintaining the system is a futile exercise. Professional evaluation and judgement is needed to determine if a coating system is capable of accepting another coat of material. All too often, the stresses developed by applying a new coating acts as a paint stripper rather than as a protective coat. The concept of encapsulation is not valid for lead paint on structural steel. Encapsulation denotes surrounding the lead on all sides. Applying a material to the top of the coating system does not encapsulate the lead as the primer is in contact with the substrate. Failures of industrial protective coatings usually occur as disbondment of the primer from the surface, undercutting by rust, or cohesive failure in the primer due to oxidation. The weak link in the coating system is at or near the steel surface. Maintenance painting is a short-term extension of coating life, with three to seven years extended life considered to be the normal range. At this point there is loss of 10 percent of the coating system or more.

The need for surface preparation presents unique situations compared to lead paint removal in other sectors. One is the volume of dust and debris that is generated. Abrasive blasting uses about 5–10 pounds of abrasive per square foot to properly remove all the contaminants. That same square foot would have only a few ounces of the lead-containing coating on it. So the amount of lead-containing debris is magnified by contributions from the abrasive. The amount of waste generated is usually expressed in tons, not pounds. The amount of dust generated during abrasive blasting is also large, as pulverization of the abrasive contributes to the dust concentration. Large dust collectors are needed to ventilate work areas to control this dust.

The critical factor in the design of a containment system is ventilation (air flow) to reduce worker exposure. Lead concentrations in the tens of thousands of micrograms per cubic meter of air have been measured in the breathing zone of abrasive blasters in well ventilated containments, and as high as 150,000 micrograms of lead per cubic meter of air in poorly designed systems. Depending on the size of containment used to isolate the work area, dust collector capacity in the 5,000–30,000 cubic feet per minute range and higher is normally needed. Smaller dust collectors do not have adequate capacity and quickly clog. Constructing containment is another unique dimension faced by people working on steel structures. Many of these structures are large or tall. Containment structures must be constructed to catch the debris and control the release of lead-containing particles. This is not a simple task, as the structures were not built with containment in mind. The lack of consistency in geometry, height, and site limitations makes each project unique.

LEAD-BASED PAINT REMOVAL CONSIDERATIONS

The industrial painting industry is still in its infancy with regards to lead paint removal. A survey of equipment manufacturers performed in the early 1980s revealed that the manufacturers were not developing equipment, tools, and devices to deal with lead paint removal as they did not see a market to justify the expense.[5]

As the industry became more aware of lead, the last decade has seen an explosion in equipment, tools, and devices to assist the contractors in performing the work. New surface preparation methods have been developed, tools have been modified, and containment devices and assemblies have become available. The effectiveness of these advances is currently being evaluated to determine effective methods and procedures.

The different methods of surface preparation available today have their advantages and disadvantages. There is no one method that has been proven to be universally accepted as the best. Local site conditions and requirements also affect the choice of surface preparation method. Following is a discussion of the different methods available to the industry. The main methods of surface cleaning or surface preparation include:

- Hand-tool and power-tool cleaning
- Water jetting
- Chemical stripping
- Wet abrasive blasting
- Blast cleaning with expendable abrasives
- Blast cleaning with recyclable abrasives

Hand-tool cleaning involves the use of scrapers, wire brushes, sandpaper, chipping hammers, putty knives, and so on to remove only loose rust, loose paint, and loose mill scale. Very little dust is generated and the quantity of debris is low. Tarps or screens can be used to isolate the work area and elaborate containments are not necessary. The intact coating must be sound and capable of accepting a repair coating. Production rates are low and so is the cost. There is the possibility of exposure of workers to lead.

Power-tool cleaning falls into two categories—cleaning to remove loose material and cleaning to bare metal. Power tools generally consist of needle guns, power wire brushes, sanding discs, grinders, descalers, and power chippers. Traditional power-tool cleaning as defined by SSPC Surface Preparation Specification Number 3 (SSPC-SP3) removes only loose material similar to hand-tool cleaning. A new SSPC specification, SSPC-SP11, requires removing all the paint, rust, and mill scale and leaving a bare metal surface. Power-tool cleaning generates a moderate amount of dust, and the volume of debris is still considered to be relatively low. Equipment costs are low. Productivity is low, but the amount of containment necessary is minimal and usually consists of ground covers and screens or tarps. Existing data show that workers will be exposed above the permissible exposure limit (PEL) for lead and therefore must be properly protected.

Power tools with vacuum attachments reduce dust and facilitate collection of the debris. Vacuum shrouded tools work well on a flat surface and different heads are available to clean common shapes such as edges and corners. However, vacuum attachments usually cannot get a proper seal on nuts, bolts, back-to-back angles, and so on. Containment needs are minimized when vacuum shrouded tools are appropriate, and exposure of workers is also minimized.

Chemical stripping involves the use of caustic or solvent-based pastes to remove the paint. The materials are applied by trowel or spray, allowed to stand, then removed by scrubbing or washing. The surface may require neutralization. The debris and flushing agents must be collected for disposal. Chemical strippers remove the existing paint only and do not remove rust or mill scale. The debris may be classified as hazardous waste for corrosivity and/or toxicity.

The equipment needed is inexpensive to moderately expensive depending on the project size and equipment needed to collect the rinsing. No dust is generated and containment requirements are moderate to low. Water washing with power washes to remove the stripper or finish from the surface may generate some airborne lead.

The amount of debris generated is moderate to large depending on the amount of water, if needed, used to flush the surface. Water discharge regulations may require collection and treatment of the water prior to disposal.

The use of chemical strippers involves other safety hazards to workers in addition to lead. Caustic strippers require proper protective equipment to avoid contact with the skin and eyes. Solvent-based strippers introduce flammability and toxicity hazards. These hazards must be properly addressed when working with these materials.

Water jetting involves the use of high-pressure water, alone, to perform surface preparation. Pressures up to 20,000 pounds per square inch (psi) are used. Loose paint, rust, and mill scale will be removed and depending on condition of the existing coating, some or most of the coating will be removed at the higher pressures. Ultra-high-water jetting—that is, pressures above 20,000 psi—will remove tightly adherent coating, but the use of water alone will not remove mill scale and rust or produce an anchor profile in steel.

The investment in equipment is expensive, although there are contractors who already own the equipment. Productivity is moderate. No dust is generated, though water-borne aerosols containing lead particles can be generated. Containment requirements can be low to sizable, depending on the amount of water used and configuration of the structure. The debris generated is low to sizable. The amount of solids collected is low as the debris consists mainly of paint chips. But if the water must be collected and filtered before disposal, the amount of debris can be sizable. Worker exposure to airborne lead emissions is significantly reduced compared to dry abrasive blasting, and an ingestion hazard still exists.

Wet abrasive blasting methods consist of either adding abrasive to a water-jetting system or adding water to a dry abrasive blasting set-up. When adding abrasive to a water-jetting method, abrasive is added after the water jet nozzles. The purpose is to remove rust and mill scale, impart an anchor profile or roughness into the surface, and to increase productivity compared to the use of just water. The spent abrasive adds significantly to the amount of waste generated, although lower amounts of abrasive are used compared to dry abrasive blasting. However, greater effort must be expended in clean-up and collection of the damp debris.

Water injection into dry abrasive blasting set-ups is performed either external to the nozzle or internal to the nozzle using special equipment. In this situation, the amount of debris is sizable. However, the water reduces the dust significantly, hence reducing worker exposure. Also, containment does not have to be as extensive as compared to dry abrasive blasting.

Dry abrasive blast cleaning is a common method of surface preparation when the total coating system must be removed and a clean substrate is desired for the new coating system. Dry abrasive blasting can be performed with expendable abrasives such as mineral sands and slags, or recyclable abrasives such as steel grit or aluminum oxide.

Dry abrasive blasting is highly effective and produces an excellent surface for painting. It also generates substantial amounts of dust and debris. Substantial containment systems are needed to protect the environment and adequate ventilation systems are needed to protect the workers. Even with the containment and ventilation systems, exposure of workers to lead exceeds the federal Occupational Safety and Health Administration's (OSHA) PELs both inside and outside containment.

The expendable abrasives generate large quantities of dust, which includes a significant contribution from breakdown of the abrasive. The waste generated is typically found to be hazardous, although recently it has been found that certain additives to the abrasive will generate a nonhazardous waste.

Recyclable abrasives significantly reduce the quantity of waste compared to expendable abrasives. Also, since the abrasive has a lower breakdown rate, less dust is generated. The equipment to recycle the abrasive to reclaim the usable portion requires a sizable investment, although contractors who have the equipment can bid jobs competitively as waste disposal costs are significantly less. When steel is used as the abrasive, not only is the volume of waste much less compared to the expendable abrasives but most of the waste has been found to be nonhazardous so disposal costs are also less.

Abrasive blasting with either expendable abrasives or recyclable abrasives can be performed with vacuum shrouded equipment. Vacuum blasting as it is called combines environmental and worker protection previously described for vacuum shrouded power tools with the ability to achieve the high degree of surface cleanliness of traditional abrasive blasting. However, all precautions noted with power tools—such as working best on flat surfaces, the need to change shrouds on edges and corners, and the inability to get an effective seal on irregular surfaces—also apply. Productivity is limited by a small blast pattern and the need to maintain the seal between the vacuum head and surface.

In addition to traditional abrasives, equipment that uses other abrasives have been developed and are in use. These include carbon dioxide pellet blasting, sodium bicarbonate blasting, and sponge blasting.

Carbon dioxide pellet blasting uses dry ice particles as the blasting medium. The dry ice particles pulverize on impact and revert to gaseous carbon dioxide. Therefore, the only debris collected for disposal are the paint chips. Dry ice will remove loose or poorly adherent coating, but works very slowly on tightly adherent coatings. It will not remove adherent rust and mill scale. Dust generated during blasting is significantly reduced and the paint is removed more in chip form than pulverized to a dust, which minimizes worker exposure and containment requirements. Carbon dioxide is also useful for removing oil and grease from machinery and had been used in lieu of solvent cleaning. Specialized equipment is needed, and it is expensive.

Sodium bicarbonate, or baking soda, is a soft abrasive like carbon dioxide. It is used with water and can be considered as dry abrasive blasting with water injection. It is effective at removing the coating but will not affect the substrate, so heavy rust and mill scale will remain on the surface. It is also effective at removing oil and grease from machinery. Worker exposure and containment requirements for lead removal is similar to other wet methods. Sodium bicarbonate is water soluble, so the abrasive dissolves after use, thus minimizing the waste. Specialized equipment is needed and its cost is moderate.

Sponge blasting uses ground up sponges as the abrasive. The sponges alone are soft and considered to be more of a cleaning medium for removing oil, grease, and loose coating. However, sponge media are available where the sponge is formed around a more traditional, harder abrasive in which case high levels of surface preparation can be achieved. Sponge blasting is a relatively new technology and experience with this method for lead removal is limited. Initial use shows dust generation is reduced and dispersion of the debris is minimized. Containment requirements may not be as stringent as dry abrasive blasting. Specialized equipment is needed, and its cost is moderate. The abrasive can be recycled.

CONTAINMENT SYSTEMS

As can be seen from the previous discussion, surface preparation will require containment to collect debris and protect the surrounding environment. This may be as simple as ground covers for hand-tool cleaning to sophisticated containment and ventilation systems for dry abrasive blasting methods. The use of containment systems within the painting industry is relatively new technology. Recently, SSPC has published a *Guide for Containing Debris Generated During Painting Removal Operations* (SSPC—Guide 61), which presents the state of the art in this area.

The guide lists containment classifications by the major components, namely the containment system and ventilation. The factors in the containment system are the containment materials, permeability of the containment material, support structure, joints, entryway, and air make-up. The ventilation system includes input air flow, air pressure inside containment, air movement, and exhaust dust filtration.

Five classes of containment are described. The simplest is Class 5 where the requirements are minimal. For example, it allows the use of permeable material such as windscreens and partially sealed joints. There is no ventilation requirement. It is intended for use with surface preparation methods that do not generate much dust.

A Class 3 containment would most likely allow dry abrasive blasting to be performed while meeting environmental regulations and providing adequate engineering controls to protect workers. Either rigid or flexible containment materials are allowed, although they should be impermeable. Joints must be fully sealed and the entryway must consist of multiple flaps. There must be forced or natural input air flow. Air pressure inside containment must be below atmospheric pressure so air flows into containment at leaks rather than being blown out. Air movement inside containment must meet a minimum velocity and the exhaust air from the ventilation system must be filtered prior to discharge into the atmosphere.

A Class 1 containment provides the highest level of protection and is intended for use in highly sensitive areas where exposure to people or equipment cannot be tolerated when dry abrasive blasting is in progress. Added requirements above a Class 3 device include an airlock entryway, controlled air make-up, and instrumental verification of negative pressure inside containment. Obviously, the costs increase with higher levels of containment.

Engineers need to be aware of the requirements for meeting regulations while performing maintenance painting. The cost of meeting these regulations can be significant. Greater thought and planning is needed to maintain a cost-effective maintenance program. But these added costs can be insignificant compared to the fines, environmental clean-up costs, liability, and bad publicity that can result if these laws are ignored.

Notes

1. B. R. Appleman, J. A. Bruno, and R. E. Weaver. *Performance of Alternate Coatings in the Environment,* Report No. FHWA-RD-89-127. McLean, VA: Federal Highway Administration, September 1990.
2. TT-P-86H, "Federal Specification, Paint, Red-Lead-Base, Ready-Mixed," Philadelphia: Naval Publications and Forms Center, July 1987.
3. E. D. Carlson. Testimony before the Subcommittee on Health and the Environment, Committee on Energy and Commerce. Washington, DC: U.S. House of Representatives, March 3, 1993.
4. CONSAD Research Corporation. "Economic Analysis of OSHA's Interim Final Standard for Lead in Construction," final report for contract No. J-9-F-1-0011. Washington, DC: OSHA, April 1993.
5. M. K. Snyder and D. Dendersky. *Removal of Lead-Based Bridge Paints,* National Cooperative Highway Research Program Report 265. Washington, DC: Transportation Research Board, December 1983.

13

Lead in Soil

Vincent M. Coluccio, Dr.P.H.

Catherine B. Ganley

The Agency for Toxic Substances and Disease Registry (ATSDR) summarized some of the key scientific concerns for dealing with lead in the following manner:[1]

> *The indestructibility of the problem.* As an element, inorganic lead cannot be processed by current technology and destroyed. It will continue to be a potential problem in some form forever.
>
> *The relative nontransferability of the problem:* Lead cannot be easily shifted from a hazardous setting to a nonhazardous setting without some concomitant increased potential risk elsewhere. Once removed from its geologically bound form by human activities, lead poses a toxic threat for which there are no natural defense mechanisms.
>
> *The environmental accumulation factor.* Lead accumulates indefinitely in the environment so long as input continues—no matter in how small a quantity.
>
> *The human body accumulation factor.* The human body accumulates lead over the individual's active lifetime and does so even with "small" intakes from common sources. For hazards to exist, major exposures at given points in time need not occur.

ATSDR continues:

> In large measure, the more tractable part of the lead abatement effort in the United States is already underway, because the reduction of lead in gasoline, food, and drinking water is amenable to centralized control. Lead in old paint, dust, and soil, however, is pervasive and dispersed, and fundamentally different approaches to abatement will be needed.

Studies conducted over the past two decades reveal extensive zones of soil lead contamination (particularly within crowded urban centers) that serve as large reservoirs of lead exposure to generation after generation of our nation's youth. These findings have prompted development of guidelines on allowable soil lead levels; remediation of lead-contaminated soils; and large-scale research studies on the connection between lead in soil and children's intake and absorption of that lead.

The topic of lead in soil is relevant to all parties engaged in managing lead hazards. Individuals assessing or managing soil lead hazards know that soil lead standards vary depending on the state or locality, and that standard protocols have not been established for measuring soil lead concentrations. In the absence of standard testing protocols, a wide variety of testing protocols are in use. Lead hazard assessors must recognize that soil lead concentrations vary both vertically (with depth) and horizontally, and therefore at a given soil sampling location highly divergent soil sampling results will be obtained based on the selection of sampling protocol. Furthermore, different units of soil lead concentration are reported, typically in units of mass per unit volume (such as parts per million, ppm) or mass per unit area (such as micrograms per square meter, $\mu g/m^2$). Moreover, even where agreement can be reached on specific soil lead measurement methods and reporting values, risk assessors may become involved in disagreements on what constitutes hazardous soil lead conditions and what (if any) soil remediation strategies to employ.

Contractors can reduce liabilities by recognizing that the public is becoming keenly aware of the potential health hazards associated with soil lead contamination caused by conventional renovation, demolition, painting, and other construction activities. Aggrieved parents and homeowners are increasingly filing lawsuits against contractors who fail to minimize, contain, and clean-up lead dusts generated by their activities.

By conducting preabatement soil lead testing, a contractor can establish that preexisting soil lead contamination was not caused by his or her activity. (Of course, this strategy can also demonstrate that soil lead levels are *higher* than preabatement levels.) It is in the contractor's interest to have an independent, qualified testing firm perform pre- and postabatement soil testing adjacent to the abatement site. In addition to quantifying soil lead levels, it is also advisable to maintain written records of visual inspections and photographic logs to document the presence of paint chips, spent abrasive blast, or other potential sources of lead that were present before commencement of the abatement project.

Another practical consideration for contractors is that high lead levels in soils surrounding residential lead-based paint abatement sites may increase the difficulty in achieving final dust cleanup levels, which are typically measured as 200 micrograms (millionths of a gram) of lead per square foot of floor surface in the interior abatement area. On such sites a greater degree of care is required to prevent soil lead dusts from being carried into the abatement site by wind or tracked in on shoes or equipment during the cleanup and final clearance phases of the abatement. Contractors are advised not to commit to achieving these clearance limits unless they are afforded control over the integrity of containment barriers and access to the abatement site during these periods.

HEALTH-BASED GUIDELINES

Based on a review of scientific literature regarding childhood exposures to lead in soil, ATSDR has reported: "A strong positive correlation is found between exposure to lead-contaminated soils and blood lead levels (PbB). Generally, the PbB levels rise 3–7 micrograms per deciliter (μg/dl) for every 1,000 ppm increase in soil or dust concentrations."[2]

Medical investigators have confirmed causal associations between soil lead and childhood lead exposures starting at 500 ppm. A variety of factors such as children's access to soil, their behavior patterns, and the presence of vegetative ground cover will influence this relationship.

The weight of the scientific evidence has prompted wide acceptance of the range of 500 to 1,000 ppm as an action level for soil lead. The U.S. Environmental Protection Agency (EPA) includes this range of action level as a factor in identifying hazardous waste sites that are eligible for federal clean-up monies.[3] Since some states and localities have adopted even lower soil lead cleanup and action criteria (New Jersey, for example, proposes a residential direct contact soil lead cleanup criteria of 100 ppm),[4] state and local agencies should be consulted for the applicable standards or guidelines.

The EPA-funded studies are examining the costs and benefits of soil lead remediation in residential neighborhoods, with a focus on determining whether remediation results in a decrease of blood lead levels among children living in the remediation area. In the meanwhile, contractors are advised to take all necessary precautions to prevent the spread of lead dusts and paint chips from their operations, particularly when working in areas occupied or frequented by young children.

SOURCES OF SOIL LEAD

Various estimates are reported for naturally occurring soil lead levels, ranging from under 50 ppm to 150 ppm, depending on the region in question. The following is a discussion of two major sources of lead that can result in localized soil lead "hot spots" well in excess of background levels: lead-based paints and leaded gasoline. Other sources of soil lead contamination are discussed in Chapter 1.

Lead-based Paints

Lead-based paints contribute to soil lead contamination in a number of ways, some obvious and some not. The chalking and gradual erosion of exterior lead-based paint contribute fine lead dusts to soils surrounding painted structures, while outright peeling and flaking of the paint film will litter the ground with discrete paint chips. Paint chips are also subject to gradual erosion as well as being crushed into smaller particles, such as by pedestrian or motor vehicle traffic.

Stripping of lead-based paint from structures, particularly by sanding and abrasive blasting methods without containment, can readily produce localized hot spots of soil levels in the thousands and tens of thousands of ppm.

Rapid changes in ambient temperature causes steel bridges (and other steel structures) to expand when hot and contract when cold; the leaded coatings on these structures may expand at different rates than the steel and thereby separate from the steel and fall to the ground as discrete chips. Other factors encouraging shedding of paint chips from steel bridges include mechanical vibrations (caused by vehicular traffic), use of corrosive salts as deicing compounds, and improper applications and aging of leaded coatings and primers.

Lead-based paints can also contribute to soil lead levels through weathering and decomposition of wood and other lead painted debris, and from past spillage onto soils of leaded paints and solvents used in the cleansing of painting equipment.

Leaded Gasoline

While the phase-out of leaded gasoline has been successful in reducing airborne emissions of lead, the deposition of this lead is a major contributor of soil lead contamination, particularly along heavily traveled roads and in urban centers.

Research by Mielke et al.[5] on 422 inner-city vegetable gardens in Baltimore, Maryland, showed high soil lead levels (with a maximum level of 10,900 ppm) relative to soils in surrounding areas, despite the fact that 90 percent of Baltimore's inner-city buildings are characterized as unpainted brick structures.

In Baltimore as elsewhere, the soil lead profile will reflect historical land use activities and an array of possible lead inputs including industrial emissions, incinerators, pesticides, paints, debris dumping, and so on. But Mielke et al. point out that vehicular emissions account for the elevated lead levels in gardens near unpainted structures in the inner-city of Baltimore. Airborne lead particles may fall out onto soil; they may also collect on the sides of buildings and be washed into the soils by precipitation. The researchers postulate that Baltimore is fundamentally similar to other large urban centers and expect the soil lead distribution of all large cities to be similar. Other studies confirm the elevation in urban soil lead levels relative to surrounding suburban and rural areas.

FATE OF SOIL LEAD

Major factors affecting the fate of lead that is deposited onto soil include (1)soil pH (a measure of soil acidity or alkylinity), (2) the organic matter content of the soil, and (3) cation exchange capacity (the presence of other elements that have the potential to bind with lead).

Lead readily binds to organic matter such as leaf litter and therefore tends to accumulate in the upper layers of soil. Soils with surface contamination that are disturbed or turned under may become contaminated down to far greater depths.

The downward leaching of lead through soil is very slow in most environments, however, certain conditions will induce leaching. These include a low pH of the leaching solution (e.g., acid rain), excessive lead concentrations, and the presence in the soil of materials that bind with the lead and form soluble compounds. Lead may also mobilize from soil when lead-bearing soil particles are carried away from their original site of deposition by storm water, wind, or machinery.

In soils with pH of 5 and with at least 5 percent organic matter, atmospheric lead is retained in the upper 2–5 centimeters of undisturbed soil.[6]

SOIL SAMPLING

The overall goal of soil sampling is to characterize lead concentration levels in soils and determine if remediation or disposal is necessary based on its potential impact to human and animal populations and the environment. To accomplish this goal, a proper sampling strategy is used to provide data that may pinpoint potential hazards existing in a particular geographic area. Generally, testing will commence in areas where visible paint chips or previous knowledge of lead contamination in an area exists. Particular attention should be given to those areas where there is potential for public exposure.

Soil sampling will also provide information pertinent to the assessment of the extent of lead contamination and may help determine its source. Sampling may also be used to establish pre- and postabatement lead levels to ascertain whether abatements have been properly conducted. Once the abatement has been completed, postabatement testing can be conducted and the difference between the two results should indicate whether proper abatement procedures were implemented or whether additional clean up is required.

Planning Sampling Strategies

Sampling protocols are planned based upon the objective of a particular study. Sampling protocols will vary according to the data requirements and the complexity of the sampling sites. The following are major classifications of sampling approaches:[7]

- *Judgmental sampling* involves the fewest number of samples and incurs the highest sampling bias. Sampling sites are chosen on the basis of known history of the sampling site, visual assessment, and technical judgment.
- *Systematic sampling* requires more samples than judgmental sampling but incurs less bias. Sampling sites are usually located along a consistent grid or pattern.

- *Random sampling* requires the largest number of samples and has the least bias. Sampling sites are chosen on the basis of simple random selection.

The following illustrates an example of a judgmental sampling approach: A 55-gallon drum of hazardous waste bursts open and spills from a storage site in an isolated area. The analytical data objective is to acquire immediate information regarding the hazard, so the sampler employs judgmental soil sampling in the immediate vicinity of the spill. A judgmental sampling plan would specify the collection of the fewest number of samples to best represent the known level of contamination immediately attributable to the spill.

Using the same 55-gallon drum scenario, systematic sampling would indiciate data representative of the overall affected area as an average. An established grid of equal areas and number of spaces would be placed over the area of the spill, and samples would be collected from each grid at the same location as the center or at equal distances of measurement within each grid. Delineating the extent of contamination resulting from the drum spill may also be assessed by use of this sampling technique, but would require a larger number of samples, thus decreasing the bias of this technique.

Random sampling of the same spill area would mean selecting inconsistent sampling locations without any bias to the particular area such as is done using the judgmental technique. An example of this may mean selecting various sampling locations from areas immediately adjacent to the spill and at random locations moving away from the spill using an unstructured or inconsistent grid pattern, that is, without a consistent point in the grid such as a center point. This type of sampling approach eliminates personal bias from the data collected.

In conjunction with all three types of soil sampling examples, control or background soil samples are usually collected for comparative review of analytical data attributed to a contamination episode identified at a particular location.

There are additional simple combinations of the three major sample approaches, such as stratified (judgmental) random, systematic random, and systematic judgmental. Further variations can be found among the three primary approaches and the three combinations: The systematic grid may be square or triangular, or the layout of the site may dictate the necessary grid arrangement. Judgmental samples may be collected for research studies based upon prior knowledge about the site in order to obtain as much useful data as possible. Random sampling is generally recommended for legal cases because personal bias is not present.

Combinations of the sampling approaches presented here generally allow for the most realistic sampling program to account for unforeseen obstacles or other uncontrollable circumstances. The sampling plan should be flexible to allow for unanticipated difficulties, such as accessibility to site sampling locations, that may require adjustments in the field. All adjustments, on-site changes, and deviations from a planned sampling protocol should be completely documented.

Soil Lead Testing Protocols

Soil lead testing protocols are specifically developed to characterize the distribution pattern of soil lead levels in defined areas. The protocols are generally developed using the three sampling approaches just discussed. Two examples of soil lead testing protocols are (1) line source or transect sampling and (2) grid area sampling. Composite sampling—for example, the collection of five subsamples (one plug from the center and one from each of the four corners) that are combined, homogenized, and submitted to a laboratory as one sample—is performed as part of both protocols. Both protocols have advantages and disadvantages, and the use of one or the other depends on the information required regarding each unique environmental study.

Transect Sampling

Transect sampling is used when the source of the contamination is identified and the sampler wishes to establish contamination levels directly attributable to a particular source. For example, a preliminary survey and sampling of surface soil lead levels beneath and adjacent to a lead painted bridge will characterize the pattern of soil lead concentrations in a defined area and demonstrate the fluctuations in these soil lead levels with increasing distances of equal length from the centerline of the bridge. Transect sampling is generally used to determine locations that reflect the greatest potential for exposure to lead in soil. In other words, this sampling technique enables the sampler to identify isolated areas of high contamination frequently referred to as hot spots. For this type of sampling, each composite sample is collected from a 12 × 12 in. square area.

One advantage to utilizing the transect line sampling technique is that one may potentially identify specific, isolated areas that have elevated soil lead concentrations. Another advantage is the ability to determine where the highest levels of lead are found and where they decrease when moving away from an identified lead source. The primary disadvantage of this technique is that it will not provide data representing an average of soil lead in a broader area of concern.

Grid Area Sampling

The grid area sampling configuration enables the sampler to obtain average lead concentration information representative of large areas, which is its biggest advantage over transect sampling. This, in turn, provides data that gives an overall assessment of the average soil lead levels within that subject area and with respect to migration considerations. For example, a composite sample collected from a 50 × 50 sq. ft. grid will present the average soil lead level within a 2,500 sq. ft. area. Surface soil from five locations within the grid make up each sample and the composite sampling specifically includes subsample collection from the middle of the grid, where diagonals of an X plot cross, and from the four point half-way between the middle and the four corners. The five subsamples are homogeneously combined for analysis.

Data Quality Objectives

Data quality objectives (DQOs) are qualitative and quantitative statements that provide critical definitions of confidence required in drawing conclusions from collected project data. The end use of collected data and analytical results is planned based upon established DQOs. These are established prior to data collection and provide a guideline outlining data collection activities, such as sufficient data; usable data; sampling arrangements; required equipment; experienced personnel; a listing of analytes, laboratory methodology and desired level of detection (LOD); sampling methodology and required quality assurance/quality control (QA/QC) protocols; sampling protocol; data analyses; and sufficient number of samples.

The uniqueness of each site under environmental study for any purpose does not allow for the development of a generic set of DQOs. The individual preparing the DQOs should take advantage of previous experience and data collected from sites with similar characteristics.

The level of detail and data quality required for each case or study will be determined based on intended data use. The DQO process contributes greatly to a detailed sampling and analysis plan that specifies the chosen sampling and analysis options and supports statements concerning the confidence in decisions made during sampling programs.

All sampling programs and investigative activities should be performed and documented in a way that ensures sufficient data of a known quantity and quality are collected to support decisions regarding remedial activities, if they are found to be necessary.

SOIL ABATEMENT

Factors to consider in choosing an appropriate soil abatement strategy include the concentration and depth of lead contamination in the soil; land use activities; surface drainage conditions; availability and costs of disposal options; and regulatory requirements.

One common strategy for soil lead abatement involves removal and replacement, whereby the top layer of contaminated soil is excavated (specifications generally range from 3 in. to 24 in. in depth) and disposed of in an appropriate disposal facility. The site is then regraded with loam or soil and revegetated. Revegetation involves resodding and seeding to stabilize the site, minimize erosion, and control dust. Ongoing maintenance of vegetative cover is required, particularly where lead contaminated soil remains under groundcover. With all replacement options, materials that replace lead contaminated soil should first be tested to assure they are not also lead contaminated.

Other common strategies for managing soil lead hazards include:

- Covering soil with uncontaminated top soil, and revegetation
- Covering soil with materials ranging from asphalt and concrete to a simple covering of wood chips or stones
- Extracting lead from the soil
- Removal and decontamination of contaminated soil, then returning the decontaminated soil onsite, and revegetation
- Roto-tilling (mixing) the contaminated soil with uncontaminated soil, and revegetation
- Fencing off a contaminated area to create inaccessibility
- Using large-sweep HEPA vacuum to pick up visible paint chips or other surface materials which may be lead contaminated
- Solidificaiton and stabilization of soil

In addition to reducing soil lead concentrations, an effective soil lead abatement strategy should include provisions to protect abatement workers, nearby residents, and community properties from lead exposure during abatement activities; locate uncontaminated replacement soils; identify disposal sites; prevent offsite runoff of lead contaminated soil caused by erosion and flooding; and prevent contamination of replacement soil.

Contractors are required to maintain a written health and safety plan for lead abatement work. The plan identifies key individuals responsible for maintaining site safety and provides detailed site-specific safety procedures to be followed. Some of the key elements of a health and safety plan are:

- Safety and health hazard analysis for site tasks and operations
- Employee safety and health training
- Selection and maintenance of personnel protection equipment
- Medical surveillance
- Environmental and personal monitoring strategy and protocol
- Site control measures
- Personnel and equipment decontamination procedures
- Emergency response plan
- Confined space entry procedures
- Spill containment program
- Organizational structure for safety and health

SOIL SAMPLING CASE STUDY

ATC Environmental's extensive sampling of soils beneath bridge and water tower abrasive blast sites reveals soil lead concentration profiles similar to the plot shown in Figure 13-1. Shown on the plot are two sets of double lines, representing the east and west outer edges of a bridge roadway, and the soil lead concentrations beneath the bridge both before and after lead-based paint was sandblasted off of portions of the bridge.

Purpose of Sampling

The purpose of the sampling in this case study was to characterize soil lead concentrations beneath and adjacent to a bridge, which is coated with lead-based paint and lead chromate primer, and to assess the impact of abrasive blasting operations on soil lead levels. A judgmental sampling scheme was chosen and consisted of line source pattern sampling, which is used when the contamination source is thought to be linear. Soil samples were obtained along linear transects extending perpendicular from the bridge in order to correlate soil lead concentrations with increasing distances from the bridge structure.

Sampling Methods

The transect represented in Figure 13-1 extends 180 ft. east and 180 ft. west of the bridge. Composite soil samples were collected at 10-ft. intervals along the

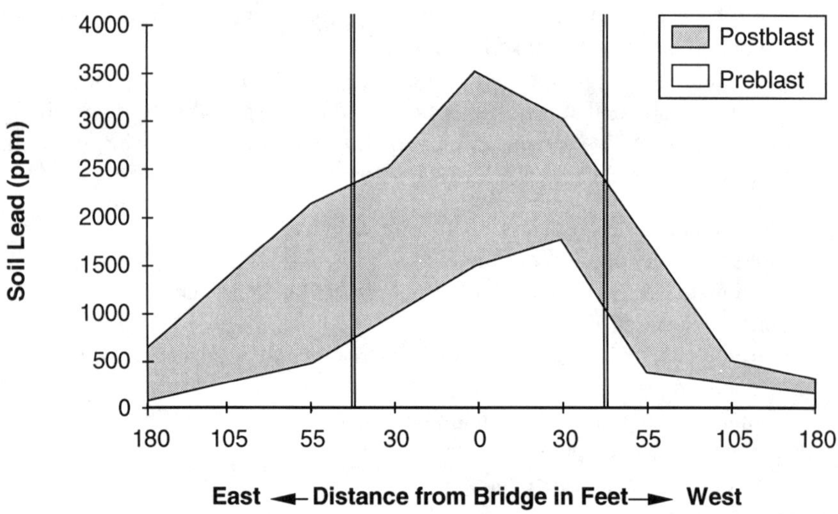

FIGURE 13-1. Soil lead concentrations versus distance from bridge.

transect. At each sampling location a 12 × 12 in. template was placed on the ground and five soil plugs were obtained, one plug from the center and one from each of the four corners of the template. Each of the five plugs from a sampling location were thoroughly mixed together into a composite sample. Samples were obtained from the upper 2 cm of soil, since this layer is of greatest concern from a risk assessment standpoint and will be most immediately impacted by abrasive blasting operations.

Small marker flags were placed at each sampling location along the transect and a photograph was taken of the transect. Samples were obtained using a stainless steel soil sampling device. Prior to taking each soil sample, the field technician donned a pair of new, clean disposable gloves, and rinsed (with dilute nitric acid) and wiped the soil sampling device with a clean towelette. Vegetation or other matter that covered a soil sampling location was brushed aside prior to sample collection, and notation was made of the nature and amount of this material in the sampling area.

Each sample was then sealed and the sample identification number written on the container. The following information was then recorded by the field technician: sample location; sampling date; whether the sampling location was tilled; presence or absence of vegetative cover; absence or presence of visible sandblast debris or paint chips; presence of painted structures near sampling locations; and other relevant data.

QA/QC Procedures

During sampling rounds, field blank-rinse samples of the stainless steel sampling device were obtained. The device was rinsed with a decontamination rinse consisting of dilute (5–10%) nitric acid (HNO_3) solution; this decontamination rinse solution was then collected in a dedicated plastic container that was sealed and transported to ATC Environmental's laboratory for proper disposal.

Following the decontamination rinse, the sampling device was then wiped clean with paper towels and rinsed with distilled deionized water. Rinse waters were collected in 250 ml precleaned collection vials that contained nitric acid preservative, and they were analyzed for lead content expressed in the laboratory report as micrograms of lead per liter of water (μg/l).

In addition, 10 percent of the composite soil samples were submitted as "split" samples to a separate certified laboratory.

Chain of Custody Procedures

Chain of custody was initiated at the time of sample collection and maintained throughout the sample handling process. Chain of custody forms and procedures were adhered to. The forms contained information regarding:

- Sampler's signature. The person printed his or her name, as a signature may be illegible.
- Name of the accredited laboratory performing lead analysis.
- Location where sample was obtained, date, time, sample number, type of sample.
- Notations that at least 200 g of soil were obtained for each sample. The soil samples were then cooled and stored to a temperature of 4°C. The samples in the laboratory were continually stored at 4°C.
- Adequate sign-off boxes for the person's signature and printed name. The sign-off boxes showed by whom the sample was relinquished and who received the sample, along with date and time.

The shipment container was then taped shut with evidence tape at all access points. The soil samples were packed in a cooler container provided by the laboratory, carefully sealed, and delivered to the laboratory.

Laboratory Analytical Procedures

All samples were analyzed for total lead in accordance with an EPA approved method, SW846 Method 6010, utilizing inductively coupled argon plasma for analyzing soil samples for lead.

Results

Preblast Sampling Results

As shown on Figure 13-1 the maximum preblasting soil lead concentration measured beneath the bridge was 1,600 ppm, well above EPA's guideline limit of 500 ppm for soil on which children play. This lead is an accumulation of fallout from prior abrasive blasting activities, lead exhausted from vehicles using leaded gasoline, and other sources described in the text.

Postblast Sampling Results

For several days following the preblast sampling, portions of the bridge structure were abrasively blasted within conventional containment barriers, consisting of tarps suspended on the underside of the bridge. During abrasive blasting, large volumes of fine dusts and coarse particles were emitted from the containment.

Each postblast soil sample was obtained in the same manner as, and within several inches of, the corresponding preblast sample. Postblast soil lead levels depicted in Figure 13-1 are as much as 2,000 ppm greater than corresponding preblast samples. Plotting of postblasting soil chromium levels (not shown) shows configurations that are nearly identical in profile, although lesser in magnitude, to the soil lead profile. This soil chromium contamination results from the release of lead chromate primer during sandblasting.

Significance of Findings

These data are consistent with ATC Environmental's findings at other steel structure abrasive blast locations. The data suggest that for the particular sites studied in which conventional containment (tarps) was used, (1) measurable impacts of lead-based paint abrasive blasting on soil lead levels occurs within an approximately 200-ft. radius of the blast site, and (2) conventional containment consisting of loose fitting tarps is not adequate for preventing soil lead impacts that exceed EPA's 500 ppm guidance limit.

Notes

1. Agency for Toxic Substance and Disease Registry (ATSDR). *The Nature and Extent of Lead Poisoning in the United States: A Report to Congress.* July 1988, p. 15–16.
2. ATSDR. *Analysis Paper: Impact of Lead-Contamionated Soil on Public Health.* May 1992.
3. U.S. Environmental Protection Agency. Interim Guidance on Establishing Soil Lead Cleanup Levels at Superfund Sites, OSWER Directive #9355.4–02. September 7, 1989.
4. New Jersey DEPE. *Proposed Rule on Soil Cleanup Standards for Contaminated Sites, N.J.A.C.* 7:26D. February 3, 1993.
5. H. W. Mielke, J. C. Anderson, K. J. Berry, P. W. Mielke, R. L. Chaney, and K. Leech. Lead concentrations in inner-city soils as a factor in the child lead problem. *AJPH* 73(12):1366–1369, December 1983.
6. Environmental Protection Agency. *Air Quality Criteria for Lead, June 1986* and *Addendum, September 1986,* EPA 600/8-83-018F. Research Triangle Park, NC, 1986.
7. Lawrence Keith. Planning and sampling protocols, in *Environmental Sampling and Analysis, A Practical Guide.* Chelsea, MI: Lewis Publishers, 1991.

14

Worker Safety and Health Regulations: OSHA Interim Final Rule on Lead Exposure in Construction

David Chambers

On June 3, 1993, a new Occupational Safety and Health Administration (OSHA) interim standard for lead in the construction industry (29 *CFR* 1926.62) went into effect. This standard drastically changed the previous construction standard (29 *CFR* 1926.55) and brought it closer in line with OSHA's *General Industry Standard.* Prior to the promulgation of this new standard, employers were forced, either by concern for their employees or fear of litigation, to follow the *General Industry Standard* (29 *CFR* 1910.1025; see Appendix E), which did not technically apply to construction-related activities. This chapter examines the key points found in 29 *CFR* 1926.62.

SCOPE AND APPLICATION: 1926.62 (a)

The interim final lead standard applies to all occupational exposure to lead in all construction work in which lead, in any amount, is present. Construction work is defined as work for construction, alteration and/or repair, including painting and decorating. It also includes the following:

1. Demolition or salvage of structures where lead or materials containing lead are present
2. Removal or encapsulation of materials containing lead
3. New construction, alteration, repair, or renovation of structures, substrates, or portions thereof that contain lead or materials containing lead
4. Installation of products containing lead
5. Lead contamination/emergency cleanup

6. Transportation, disposal, storage, or containment of lead or materials containing lead during construction activities
7. Maintenance operations associated with construction activities

PERMISSIBLE EXPOSURE LIMIT: 1926.62 (c)

Prior to the promulgation of this standard, the permissible exposure limit (PEL) for lead in the air in the construction industry was 200 $\mu g/m^3$—a level four times higher than the standard set for the general industry and universally accepted as far too high for workers in the construction industry. The interim final standard rectified this gap in worker protection standards by lowering the PEL for the construction industry to 50 $\mu g/m^3$, averaged over an 8-hour period. In instances where the employee is working for a greater than 8-hour period, the formula for determining an employee's allowable exposure is as follows: 400 divided by hours worked in the day.

In conjunction with the lowering of the PEL in the construction standard to 50 $\mu g/m^3$, OSHA also set the action level at 30 $\mu g/m^3$, which is also in line with the *General Industry Standard.*

EXPOSURE ASSESSMENT: 1926.62 (d)

If lead is present in any quantity in the workplace, an employer is required to make an initial determination of whether any employee's exposure to lead exceeds the action level. There are a number of ways this initial determination may be made, the most obvious being air monitoring. However, if an employer can produce objective data that proves an employee is not being exposed to lead above the action level, then air monitoring would not be required. If air monitoring results are to be used as objective data, then this air monitoring must have been performed within the last 12 months and be representative of employee work tasks and exposure conditions and meet the requirements for accuracy specified in 1926.62. The key to using objective data to prove low employee exposures to lead is that it must be directly applicable to the current activities being performed by the employees.

If air monitoring demonstrates the likelihood that employees may be exposed to air levels that may exceed the action level (without regard to respirators), then the employer must set up an air monitoring program that is representative of employee lead exposure. As is often the case with OSHA, the concern is not how many employees are sampled but whether the sampling is representative of the work being performed. For example, an employer could air sample 30 percent of the work force but the individuals sampled are clerical or office personnel. If no one out in the field is sampled—the individuals actually performing lead-related activities—then the air sampling program is nonrepresentative of the work and does not meet the OSHA standard.

Once an individual has been sampled, the employer must notify that person of the results within five working days after receipt of the analytical results. If the results exceed the PEL, the employer must notify the employee in writing of this fact and describe actions that have been or will be taken to reduce the employee's exposure.

If an employee's exposure level is at or above the action level but below the PEL, that individual's exposure must be reexamined every six months. Air sampling can be discontinued if two consecutive measurements, taken at least seven days apart, are at or below the action level. If an employee's exposure level exceeds the PEL, subsequent air sampling must take place every 3 months instead of every 6 months. The air sampling may be discontinued when two consecutive samples, taken at least 7 days apart, indicate that the employee's exposure is below the PEL. If the results indicate that an employee's exposure is below the PEL but above the action level, then air sampling continues every 6 months, as described previously.

Perhaps the most important aspect of an OSHA air monitoring program is the concern for a change in work conditions, equipment, control measures, or a new type of job function. In these instances an employer is obligated to begin air sampling again to establish exposure levels under the new conditions. The same aforementioned air sampling protocols apply to this new round of air sampling.

The preamble to the OSHA construction standard describes various construction activities and associated airborne lead concentrations. Table 14-1 lists several of these construction activities to illustrate the vast differences in potential exposures.

Through OSHA, the government has also compiled a list of activities that will require interim worker protection until objective data or air sampling results prove otherwise. These activities include: abrasive blasting, welding/cutting/burning of steel structures, lead burning, manual scraping/sanding, manual demolition of structures, heat gun applications, use of lead-containing mortar, moving or removing an abrasive blasting containment, power tool cleaning, rivet busting, and clean-up activities where dry expendable abrasives are used. The employer is required by OSHA to assume a worker will be exposed to lead in excess of the PEL when conducting any of these activities. Therefore the employer is obligated to

TABLE 14-1. Potential Lead Exposures from Selected Construction Activities Involving Lead-based Paint

Construction Activity	Lead Exposure (μg/m^3)
Open abrasive blasting in full containment	37,300
Open abrasive blasting	23,680
Welding/cutting/burning	973
Scraping	101
Chemical stripping	15
Encapsulation	4

Source: Preamble to 29 *CFR* 1926.62: OSHA Interim Final Rule on Lead Exposure in Construction.

protect workers until air monitoring (or previously established objective data) establishes that protection is unnecessary. Protection will include respirators and may extend to other levels of personal protective equipment.

METHODS OF COMPLIANCE: 1926.62 (e)

According to OSHA, an employer's main responsibility is to ensure that no employee is exposed to lead in excess of the PEL as an 8-hour time weighted average. The standard requires employers to implement workplace controls to the extent feasible to reduce employee lead exposure. If it is not possible to reduce lead exposures to below the PEL, then an employer must supplement any workplace and/or engineering controls with the appropriate respiratory protection.

Prior to the start of any work activities that may expose employees to lead above the PEL, an employer must have a written compliance program. Elements of the plan include:

1. Description of lead emitting activity
2. Description of the means employed to achieve compliance with the OSHA standard including a description of any engineering controls used and their effectiveness
3. Air monitoring data
4. Details on the purchase of protective equipment, high efficiency particle air (HEPA) filtered vacuums, etc.
5. Details of any administrative controls used to control employee exposures—i.e., employee job rotation
6. Description of the arrangements made with other contractors working at the job site to inform their employees of the hazards of lead

A major element of the compliance program is the provision for frequent and regular inspections of the job site, equipment, and materials by a competent person. The compliance program must be updated every 6 months and be made available to OSHA, affected employees and/or the employees' representative.

RESPIRATORY PROTECTION: 1926.62 (f)

Respiratory protection must be provided to employees at no cost to them. The use of respirators is mandated under four conditions: (1) Whenever exposure levels exceed the PEL; (2) in work situations where engineering controls and work practices will not be sufficient to reduce exposures to at or below the PEL; (3) whenever an employee chooses to use a respirator, regardless of air levels; and (4) as an interim protection until air monitoring results and/or objective data prove that exposures are at or below the action level.

All respirators must be approved by the National Institute of Occupational Safety and Health (NIOSH) and the Mine Safety Health Administration (MSHA) and must provide protection against lead dust, fume, and mist. An employee has the right to request a powered, air purifying respirator (PAPR) in lieu of any negative pressure respirator as long as it will provide adequate protection to the employee.

For any employee using a negative pressure respirator, the employer must ensure that it fits the individual properly. To verify a proper fit, the employer may perform a qualitative or quantitative test at the time of initial fitting and every six months thereafter for those employees using negative pressure respirators. If an employer chooses to use the simpler qualitative test to fit a respirator, then only half-mask respirators may be used. The qualitative test is *only* valid for half-mask respirators. If at any time during the fit-test, or while wearing a respirator, an employee experiences difficulty in breathing, the employer must offer a medical examination to determine whether or not that employee can wear the respirator.

Employees who use respirators are permitted to change filters whenever there is an increased breathing resistance and it is the duty of the employer to provide enough replacement filters for this purpose. Also, the employees are allowed to leave the work area as often as needed to wash their face and respirator facepiece to prevent skin irritation.

PROTECTIVE WORK CLOTHING AND EQUIPMENT: 1926.62 (g)

Protective clothing must be provided to employees, at no cost to them, appropriate to the level of hazard they will or may encounter. In environments that exceed the PEL, the employer must ensure that protective clothing is not worn home. The standard states that clean clothing must be provided weekly to employees who are exposed above the PEL and in environments that exceed 200 $\mu g/m^3$, the employer must provide a daily change of clothing. It is the employer's responsibility to provide a clean change area for the removal of protective clothing.

Cleaning, laundering, or disposal of protective clothing must be provided by the employer. If protective clothing is to be laundered, the employer must notify the individual or company that provides this service of the potentially harmful effects of lead exposure. While storing contaminated clothing, the employer must provide a container labeled as follows:

> **Caution:** Clothing contaminated with lead. Do not remove dust by blowing or shaking. Dispose of lead-contaminated wash water in accordance with applicable local, state, or federal regulations.

HOUSEKEEPING: 1926.62 (h)

The goal of this section of the regulation is to maintain the work area as free as practical of accumulations of lead. This can be accomplished by vacuums equipped with HEPA filters, wet mopping or sponging, or shoveling and brushing. If shoveling or brushing is used, the employer must establish that these methods are more effective than HEPA vacuuming or mopping. At no time may compressed air be used to remove lead from surfaces unless the compressed air is used in conjunction with ventilation specifically designed to capture airborne dust created by the compressed air system.

HYGIENE FACILITIES: 1926.62 (i)

A clean change area must be provided to employees whose exposures are above the PEL and for all employees performing the activities listed in 1926.62 (d). The change area must have separate storage facilities for contaminated clothing, labeled as per 1926.62 (g). No employee is allowed to leave the change area wearing protective clothing required to be worn during the work shift.

If feasible, the employer should provide shower facilities to those employees exposed to lead above the PEL. The showers must be maintained in a clean condition and soap and towels are to be provided by the employer. If showers are not provided, a hand washing facility must be provided in accordance with 1926.51 (f) and the employer must assure that employees will wash their face and hands at the end of the work shift.

In areas where employees are exposed to lead above the PEL, food, beverages, and tobacco products may not be present or consumed, and cosmetics may not be applied. A lunchroom or eating area as free of lead dust as possible must be provided for employees. No employee may eat or drink in protective clothing unless that clothing has been HEPA vacuumed or other methods employed to remove surface contamination.

MEDICAL SURVEILLANCE: 1926.62 (j)

A medical surveillance program is one of the most important aspects of the construction standard and is a significant part of the standard's comprehensive approach to the prevention of lead-related disease. However, OSHA acknowledges that the best medical surveillance program and worker protection standards may not protect workers who have high body burdens of lead accumulated over the years or who have additional uncontrolled sources of lead exposure beyond the workplace. There are also individuals who absorb lead at a greater rate than the normal person and workers who have certain medical conditions that may be aggravated by lead exposure. For those individuals, the medical surveillance program may not be as effective.

The standard divides the medical surveillance program into two parts—biological monitoring and medical examinations. It is not mandatory for an employee to participate in the program, but it is required that an employer offer the program to an employee at no cost to the worker and at a reasonable time and place. The program must be performed by or under the supervision of a licensed physician. The need for a medical surveillance program is triggered by the results of the air monitoring program.

Biological Monitoring

For those employees exposed to an airborne concentration of lead—on any one day—that is at or above the action level, the initial medical surveillance program will include biological monitoring, that is, blood sampling and analysis for lead and zinc protoporphyrin levels. If the results indicate the employee's blood lead level is at or above 40 μg/dl, the employer must continue biological monitoring until two consecutive tests indicate the blood lead level is below 40 μg/dl.

If an employee's exposure is at or above the action level for more than 30 days per year, a biological monitoring program must be implemented. Biological monitoring must take place every two months for the first six months of employment and then every 6 months thereafter. However, should an employee's blood lead level reach 40 μg/dl or above at any time during the monitoring period, the frequency of tests must increase to every two months. Once two consecutive blood tests indicate a blood lead level below 40 μg/dl, the monitoring will continue every 6 months.

Two additional biological monitoring requirements are triggered by blood lead level results: (1) For each worker whose blood lead level requires immediate medical removal, a monthly test must be performed and (2) For each worker who requires temporary medical removal, a follow-up test must be conducted within 2 weeks. (See the criterion for medical removal in the next section.)

Notifying employees of their test results is an essential element in the standard. Within five working days after receipt of blood test results, the employer must notify the worker in writing of their blood lead level. Employers must notify workers whose blood lead level exceeds 40 μg/dl of when medical removal may be necessary and of the medical removal protection benefits granted under the standard.

Medical Examinations

Prior to conducting a medical examination, an employer is obligated to inform the physician that any results of the exam that are unrelated to lead exposure may not be revealed to the employer. In addition, the employer must advise the employee of any medical condition that may dictate further medical treatment. An employer's obligation to provide a full medical surveillance program to an employee, which includes medical examinations, is triggered by a number of different occurrences:

- Prior to an employee's assignment to a work area, the area shows airborne lead concentrations that meet or exceed the action level.
- An employee is exposed to lead above the action level for more than 30 days per year and has had a blood lead level at or above 40 μg/dl within he past 12 months.
- An employer is notified by an employee that he or she has developed signs or symptoms commonly associated with lead intoxication.
- The employee desires medical advice regarding lead exposures and the ability to procreate a healthy child.
- The employee demonstrates difficulty in breathing while being fit for a respirator or while wearing a respirator.
- A medical examination is required because employees have been removed from exposure due to risk of sustaining material impairment to their health.

The required examination includes a work and medical history along with a general physical. The employee's pulmonary function status should be evaluated if respiratory protection will be used. A routine urinalysis, blood analysis, and any other tests deemed necessary by the physician complete the medical surveillance program.

An employer has the obligation to notify an employee of his or her right to seek a second medical opinion, although this notification does not have to be in writing. Should an employee feel it necessary to consult with a second physician regarding the medical examination results, he or she has 15 days to inform the employer of this wish. As OSHA mandates, the employer must pay for this second opinion. If the second medical opinion differs from the first and the differences cannot be settled between the two physicians, then a third medical opinion may be sought. This, too, must be paid for by the employer.

The employer must provide information to the physician conducting the medical examinations including: A copy of the OSHA lead construction standard; description of the employee's duties as they relate to lead exposure; the employee's known or expected exposure level to airborne concentrations of lead; description of any personal protective equipment to be used by the employee; any prior blood determinations; and all prior written medical opinions concerning the employee in the employer's possession or control. This information must be provided by the employer to the second or third physician upon the request either by the second or third physician, or by the employee.

Once the examination is completed, the physician must provide a written medical opinion to the employer that includes the blood lead results. This opinion will also include any medical conditions detected that may place the employee at an increased risk of material impairment from lead exposure. In addition, the opinion will also detail any special limitations to be placed upon the employee in regard to the wearing of a respirator.

The use of chelating agents is addressed in the standard and requires the employer to assure that it is conducted under the supervision of a licensed physician in a clinical setting. Since chelation uses very powerful drugs to remove lead from the blood, an employee must be notified in writing prior to its occurrence. At no time may any person retained, employed, supervised, or controlled by the employer engage in prophylactic chelation of any employee.

MEDICAL REMOVAL PROTECTION: 1926.62 (k)

Medical removal protection (MRP) is a means of protecting a worker when, for whatever reason or reasons, other methods of worker protection have failed. The method involves the temporary removal of a worker from his or her regular job to a place of significantly lower exposure without any loss of earnings, seniority, or other employment benefits. For up to 18 months, or for as long as the job the employee was removed from lasts, protection is provided.

An employee must be removed from lead-related activities when a periodic and follow-up blood test indicate a blood lead level at or above 50 µg/dl. In other words, if an employee has a periodic blood test and the result is a blood lead level of 50 µg/dl, the employer must provide a follow-up blood test within two weeks. If the result of the follow-up test is at or above 50 µg/dl as well, the employee must be removed from activities that may expose him or her to airborne lead levels at or above the action level.

Medical removal may also occur if a physician determines that an employee is at an increased risk of material impairment to health from exposure to lead. This determination must be made by a licensed physician in writing. An employer is obligated to follow the physician's opinion regarding the employee.

Once removed, an employee may return to lead-related work activities when two consecutive blood tests indicate a blood level at or below 40 µg/dl. In the case of removal because of a physician's written opinion, the employer must have a written statement by the physician that it is acceptable for the employee to return to lead-related work. During the removal period, the employer is obligated to extend all normal benefits to the employee. In extreme circumstances, the worker may be laid off during the recovery period (with full pay and benefits), although OSHA strongly urges an alternative route be explored.

TRAINING: 1926.62 (l)

As is usually the case with OSHA regulations, employee training is a definite requirement, but the specifics of who can teach and how many hours/days the training should take is vague. However, OSHA does set certain guidelines that must be covered in the training program. For all employees, a training program must be conducted incorporating the *OSHA Hazard Communication Standard for the Construction Industry* (1926.59). This includes information on lead hazards, material safety data sheets (MSDS), warning labels, and signs.

For all employees exposed to lead at or above the action level, the program is more clearly defined and includes the following:

The content of the OSHA Interim Lead Construction Standard and its appendices
Operations that may result in lead exposures
Purpose and description of the medical surveillance and medical removal program
Engineering and work practice controls to be used by the employee
Contents of any compliance plan in effect
Instruction to the employees that chelating agents should not be routinely used to remove lead from the body
Employee right to access records

Recording keeping of any employee training program should be maintained by the employer. The employer is also obligated to provide ready access to a copy of the OSHA standard for all affected employees. At any time OSHA may request information regarding the training program and any records in this regard must be surrendered to OSHA.

SIGNS: 1926.62 (m)

Where lead exposures are above the PEL, the employer shall post a warning sign at the entrance to that work area. The signs must be illuminated and cleaned as necessary so that the legend is clearly visible. The legend must read as follows:

WARNING
LEAD WORK AREA
POISON
NO SMOKING OR EATING

RECORDKEEPING: 1926.62 (n)

Recordkeeping under the OSHA standard is divided into four categories: exposure assessment, medical surveillance, medical removals, and objective data. Remember, any record generated by a lead-related work activity is important and should be maintained by the employer for future reference. This includes training records, fit-test records, accident reports, or other vital information. All records described in this section must be made available to affected employees, former employees, and their designated representatives, and to OSHA for examination and copying.

Exposure Assessment

Employee exposure assessment records are crucial to an employer, should there be a question regarding an employee's exposure to lead during a particular activity. According to OSHA, the records of each exposure assessment must include the following:

1. Date(s), number, duration, location and results of each sample taken and a description of the procedure used to determine representative employee exposure, where applicable
2. Description of the sampling and analytical methods used and evidence of their accuracy
3. Respiratory protection worn, if any
4. Name, Social Security number, and job classification of employee monitored *and* other employees whose exposure the measurement is intended to represent
5. Environmental variables that could affect the measurements

Medical Surveillance

Medical surveillance records must include the employee's name, Social Security number, and description of the employee's duties. If there are any air monitoring results for the employee and the results were provided to the physician, then those results become a part of the medical surveillance records. Any employee complaints related to lead exposure must be maintained in the employee's file as well as any and all copies of the physician's written opinion(s).

According to OSHA, it is up to the employer to maintain, or assure that the examining physician maintains, the following records: copy of medical examinations, description of laboratory procedures and a copy of any standards or guidelines used to interpret test results, and a copy of biological monitoring results.

Medical Removals

If an employee is removed from lead-related activities due to elevated blood-lead levels or for any reason, the employer must document that employee's name, Social Security number, date the employee was removed, and a statement indicating if the removal was due to elevated blood lead. The date the employee was returned to his or her former job status must also be included, as well as a brief explanation of how the removal was accomplished. These records must be maintained for at least the duration of the employee's employment.

Objective Data

Any objective data used to document that a specific process, operation, or activity involving lead cannot release dust or fumes in concentrations at or above the action level must be maintained by the employer for 30 years.

If an employer ceases to do business, the successor shall receive and retain all records. If there is no successor, then all records are to be sent directly to OSHA. At the expiration date of the retention period for all records, the employer must notify OSHA three months prior to any disposal of records and must turn over the records to OSHA if requested.

15

Legal Liability, Loss, Risk Management, and Insurance Considerations for Lead Hazard Control Professionals

Jackson L. Anderson, Jr., C.P.C.U.

Childhood lead poisoning represents one of the greatest preventable epidemics in our society, with an estimated 3 million children already having the elevated blood lead levels necessary for diagnosis.[1] Yet to date, public and private responses to the causes of lead poisoning have not served to reduce the hazard nor the incidence of the disease.

But this situation is changing. With the recent discovery of the serious consequences of low level lead poisoning in children, a consensus is evolving among health, housing, and environmental leaders that the new public policy objective must be directed at true, primary prevention. Emphasis should now be placed on identifying and correcting lead-based paint hazards in housing *before* a child can become poisoned.

This effort is likely to stimulate growth of a lead hazard evaluation and control industry. Indeed, the Residential Lead-Based Paint Hazard Reduction Act of 1992, known as Title X, establishes provisions for training, licensing, inspections, risk assessments and other technical criteria.

However, although it is anticipated that lead hazard evaluation and control activities will be required to carry out the broad initiatives in this federal legislation, the reality is that many property owners do not have access to funds to correct lead paint hazards in their homes. Often, such investments simply do not make any economic sense. At a larger level, the lack of funds for abatement is eclipsed by the lack of financial resources and the low level of overall investment in affordable housing.

Ultimately, it is desirous to remove all sources of lead that can create human exposures. But strict enforcement of full abatement requirements, without public subsidies, could force the abandonment of millions of low value units. The magnitude and cost of this endeavor make it a long-term goal. In the meantime, the objective will be to make all housing in the United States "lead safe."

The need for qualified lead hazard evaluation and control professionals will certainly grow. Professionals who have already entered or are contemplating entry into this field must make a number of informed business decisions, not the least of which is assessing potential legal liabilities and exposure to loss. Such assessment requires that contractors, risk assessors, inspectors, designers, and planners consider implementing a comprehensive risk management program that includes needed insurance coverages. But before such a program can be developed, the lead hazard evaluation and control professional should be aware of the potential liability and loss exposures that can affect their operations.

POTENTIAL LIABILITY AND LOSS EXPOSURES

To date, few claims or suits have been brought against lead abatement contractors or consultants. But there is a growing awareness of the danger of lead. In fact, technical, legislative, and regulatory developments over the past few years have heightened the nation's concern over and response to childhood lead poisoning.

These developments are likely to stimulate the filing of lead paint poisoning lawsuits, primarily against landlords but also against the lead hazard evaluation and control professionals who do not perform their activities in a safe manner. Because children are most susceptible to lead poisoning, special liability issues are created that can have a significant impact on the value of a lead poisoning claim.

As licensing and training requirements become more demanding, and as more suits are brought and adjudicated, the legal standard of care to which lead hazard evaluation and control professionals will be held will rise dramatically. This is why there is a need to be aware of the legal liability and loss implications that potentially affect these professionals.

There are three basic types of liability exposures that apply to lead hazard control and inspection/assessment activities: general, professional, and statutory.

General liability arises from the ownership, maintenance, and use of the business premises, *and* from onsite and offsite business activities. *Professional liability* arises from the acts, errors or omissions in rendering or failing to render tests, inspections, assessments, plans, drawings, opinions, or advice. *Statutory liability* arises from legislative or administrative law, either at the federal, state, or local level.

A further word about professional liability is in order. Historically, "professionals" have been associated with occupations requiring higher education and specialized training, such as physicians and lawyers. Professionals are also those who possess a special skill and knowledge in their field; they are usually held to a higher degree of responsibility than members of the general public. The category has expanded over the last 25 years to include accountants, architects, engineers, and others. Because lead inspectors, risk assessors, designers, and planners must employ special skills, knowledge, and training to carry out their activities, they are likely to be judged as professionals.

Legal issues against lead hazard control contractors and consultants will arise from tort or common law, contract law, and regulatory statutes and guidelines.

Tort or common law is based on the accumulated decisions of the courts that can not be decided by application of statutes or regulations. (A "tort" is a civil wrongdoing for which damages can be awarded.) *Contract law* arises from the violation or breach of a legally enforceable promise between two or more parties. *Regulatory statutes and guidelines* are the administrative laws, ordinances, and guidelines promulgated by legislatures and governmental agencies.

Each of these act alone or in concert to establish the legal standards against which contractors and consultants will be held accountable. The important point here is that the standards for lead hazard evaluation and control activities are still emerging and will affect the operations of all who enter this field.

There are certain legal duties that are owed to others by all members of society. This certainly includes lead hazard control contractors and consultants, although these duties, which are directly applicable to the legal standards of care, will continue to evolve. Generally, however, there are four legal duties that will apply.

The *duty of reasonable care* turns on the foreseeability of harm to others that arises from a contractor's or consultant's activities. The legal obligation is to protect others from such harm. The *duty to warn* requires that others be adequately advised of potential hazards that can cause injury or damage. The legal obligation here is to alert third parties to such dangers. The *duty to be informed* refers to the need to keep abreast of the latest scientific information, discoveries, procedures, products, and regulations that affect or govern one's chosen field. Failure to do so can result in liability if such knowledge would have prevented injury or damage to another person. The *duty to test* requires that one evaluate and ascertain potential hazards to which third parties can be exposed.

Liability can be imposed when there has been a finding of negligence, breach of contract, breach of express or implied warranty, strict liability, or negligence per se.

Negligence requires that there has been a breach of a specific legal duty owed to another that is the proximate cause of injury and resulting damages. *Breach of contract* is, quite simply, failure to perform as per the terms of a legally enforceable promise, that is, a contract. A finding of *breach of expressed or implied warranty* requires that all conditions of safety have not been identified and met even though one may have expressly or implicitly warranted that they have. *Strict liability* is liability that can be imposed without evidence of fault on the part of the alleged wrongdoer. (At this point, although there are two state statutes that impose strict liability against landlords for lead paint hazards that have not been abated, there are no statutes that could apply strict liability against lead hazard evaluation and control professionals.) *Negligence per se* maintains that the violation of an ordinance, statute, or regulation may be sufficient to establish liability.

POTENTIAL CLAIM COSTS

Lead hazard control contractors and consultants face a variety of potential loss exposures and claim costs that can adversely affect their operations and ability to stay in business.

Damages for injury or death, not only for lead poisoning but other construction-related accidents, can be catastrophic or serious enough to affect the ongoing financial viability of a firm. Damages for wrecked or destroyed property can also have equally devastating effects on the lead hazard evaluation and control professional. Also, the cost of investigating and litigating a claim or lawsuit can be appreciable and many times will eclipse the value of the claim, particularly when it is without merit. There are also business expenses associated with liability claims such as downtime and loss of production, increased operating costs when work must be redone, loss of reputation and work opportunities, statutory fines and penalties, inability to obtain surety bonds, and increased insurance premiums.

RISK MANAGEMENT TECHNIQUES

So how does the emerging lead hazard evaluation and control industry address these potential liability and loss exposures? Perhaps the most sound way is by developing a total "risk management" approach that recognizes and plans for such exposures.

Risk management is a decision-making process for the identification, analysis, and treatment of loss exposures. Further, it involves implementation of a program to minimize the financial costs associated with liability and loss.[2]

There are five possible risk management techniques to employ:

1. Avoidance
2. Loss control
3. Noninsurance transfer
4. Retention or self insurance
5. Insurance

Avoidance is the technique of consciously avoiding a potential liability or loss exposure. In its simplest form, it is the decision not to pursue a specific business venture or opportunity because of the chance of a severe or catastrophic loss that could seriously affect or even bankrupt the firm. For some lead hazard evaluation and control professionals, it may simply mean not going after a contract or project because of a lack of manpower, equipment or "know how," or because the work is highly risky (such as an occupied multifamily rental property.)

Loss control involves taking steps to reduce the frequency and severity of loss exposures. *Loss prevention* attempts to control loss frequency. *Loss reduction* attempts to control the severity of those losses that do occur.[3]

Lead hazard evaluation and control professionals can implement a number of loss control techniques to limit or reduce the size of losses. These include:

- Properly training all workers, supervisors, and other field technicians
- Staying abreast of the latest federal, state, and local statutes, regulations, and guidelines governing lead abatement
- Developing standard operating procedures in manual form
- Developing quality assurance/quality control protocols, again in manual form
- Closely supervising all subordinates by maintaining a clear chain of command
- Preparing clear, concise, "tight" and enforceable contracts with clients and subcontractors
- Maintaining complete and accurate inspection and risk assessment documentation and project files
- Carefully evaluating, selecting, and monitoring subcontractors
- Following a program of "good housekeeping" at the business premises and the work site
- Preparing to address any emergencies such as worker injury, fire, and other first aid situations

Noninsurance transfer is the process of transferring loss exposures to another individual or firm. The most common method is by requiring some party, usually a contractor or subcontractor, to indemnify, defend, and hold harmless the firm for liability and loss arising out of the contractor's or subcontractor's work. This is a typical provision in most construction contracts and will likely be a factor in lead hazard evaluation and control activities. Another technique that a firm can implement is to require a contractor or subcontractor to carry liability insurance that lists the prime firm and/or building owner as an "additional insured."

Retention or self-insurance means "assuming the financial burden for part or all of the consequences of a particular loss exposure."[4] This technique usually works best when expected losses are of low severity and/or highly predictable. In other cases, however, the decision to self-insure or retain the loss exposure (sometimes *all* of it) may not be an option; it may be a necessity.

The types of insurance recommended or required for lead hazard evaluation and control operations typically have deductibles, thereby involving the insured in the loss. With deductibles, the insured's insurance carrier participates in the loss from "dollar one," that is, it investigates and defends the claim even though the insured may have to contribute to any potential settlement and sometimes even defense costs. However, if the insured has a "self-insured retention," it is typically responsible for any claim or lawsuit until the potential or actual dollar exposure reaches a preestablished level. This requires not only the financial reserves to cover the loss but also the ability to handle and direct the claim (which, many times, the firm may *not* have.)

Finally, *insurance* is a key risk management technique for all individuals, businesses, and organizations. Certain insurance coverages are desirable in any construction or building renovation project, and lead hazard evaluation and control projects are no exception. Therefore, contractors, inspectors, risk assessors, designers, and planners should have applicable policies in force for the entire term of the project, from bid acceptance through completion of the work, compliance with clearance criteria, and removal of all equipment, supplies, and employees. The following coverages are recommended:

- *Commercial General Liability (CGL).* The contractor and inspector, risk assessor, designer, and planner should each carry this insurance coverage. The policy should be "broad form," written on an "occurrence" basis, to include premises and operations liability, contractual liability, independent contractors liability, and products and completed operations liability. Also, it is recommended that the policy be specifically endorsed and/or written to include coverage for lead hazard evaluation and control operations and have the so-called "pollution exclusion" eliminated or modified so that it will not apply to lead hazards, exposures, poisonings, or claims. Limits of liability of $1 million per occurrence with a $1 million policy aggregate for bodily injury and property damage are suggested. The building owner should also be named as an "additional insured" on all such policies. ("Occurrence" policies require that there be bodily injury or property damage caused by an "occurrence," typically defined in various insurance industry general liability policies as an "accident, including continuous or repeated exposure to substantially the same general harmful conditions," that occurs during the policy period. There is no restriction on when a resulting claim or suit must be made or brought against the insured.)[5]
- *Professional Liability Errors and Omissions (E&O).* The inspector, risk assessor, designer, and planner should each carry this insurance. The policy will typically be written on a "claims made" basis and cover professional services rendered in connection with inspections, risk assessments, environmental sampling, project supervision and monitoring, and specification writing. Again, it is recommended that the policy have the so-called "pollution exclusion" eliminated or modified so that it will not preclude acts and errors or omissions that result in lead hazards, exposures, poisonings, or claims. Limits of liability of $1 million per claim with a $1 million policy aggregate are suggested. ("Claims made" policies require that either bodily injury or property damage, caused by an "occurrence," must occur during the policy period *and* must result in a claim or suit first made

against the insured which is reported to the insurer during the policy period. Virtually all E&O policies available today are written on a "claims made" basis; "occurrence" coverage is not an option.)

• *Commercial Automobile Liability.* The contractor, inspector, risk assessor, designer, and planner will typically be required in most states to carry this insurance coverage for all owned, nonowned, or hired vehicles, including those in use during the term of the project. However, if any vehicles also transport lead hazardous waste, they must comply with the provisions of the Motor Carrier Act, including carrying a limit of liability of $5 million per occurrence on the policy.

• *Workers Compensation.* All lead hazard control professionals must carry workers compensation insurance that complies with all the highest statutory requirements of those states where work is to be performed. Additionally, employers liability coverage should be included with a minimum limit of liability of $1 million.

THE INSURANCE DILEMMA

Unfortunately, the aforementioned insurance is not readily available to lead hazard evaluation and control professionals. That which *is* available is often expensive and/or affords little meaningful protection. Why has the insurance industry restricted this coverage in the first place? There appears to be a number of factors that have influenced this situation:

1. The insurance industry believes there is a debate in the medical and legal communities as to the causes, symptoms, injuries, and damages associated with childhood lead poisoning.
2. At present, there is no universally accepted and mandated means for making low-income housing, in particular, lead safe.
3. Data on lead poisoning claims, suit filings, settlements, verdicts, dismissals, and other factual information is limited, making it very difficult for insurers to accurately predict future loss costs and assemble appropriate rates.
4. Increased childhood blood lead screening and the public's awareness of the causes of lead poisoning, coupled with the lowered blood lead threshold level for the diagnosis of this disease, is likely to cause an increase in lead litigation.
5. The potential for explosive verdicts in urban centers is great, given the age and underprivileged nature of many of the lead poisoned children.
6. The legal "standard of care" applicable to lead hazard evaluation and control professionals is still evolving, with no nationally recognized standard currently existing.
7. Precedents from previous unrelated insurance coverage decisions may serve to expand the scope of coverage in the liability policies that will be called upon to respond to lead poisoning claims and lawsuits.

THE INSURANCE MECHANISM EXPLAINED

To further understand the reasons for the insurance industry's reluctance to provide coverage to lead hazard evaluation and control professionals, one must first understand some fundamental principles about the insurance mechanism.

Insurance has been defined as a system in which a person, business, or organization transfers the financial consequences of loss exposures to an insurer, which then indemnifies the insured for covered losses and provides for the sharing of loss costs among all insureds.[6] An insurance contract or policy spells out all the terms, conditions, definitions, and exclusions of this transfer.

By transferring loss exposures to an insurer, insureds exchange the possibility of a large loss for a much smaller periodic payment, that is, premiums. By sharing loss costs, each insured's premium is accumulated in a loss fund, with the total cost of losses spread among all insureds. In order to determine how large a pool of funds is necessary to pay for all insureds' losses, insurance companies must be able to predict future losses and expenses. This is done by using the *law of large numbers,* a key factor in and the foundation for all insurance operations.

The law of large numbers is a mathematical principle that states that when the number of similar, independent exposure units increases, the relative accuracy of predictions about future outcomes also increases. This principle enables insurers to improve the predictability of losses by pooling a large number of similar but independent exposure units. Increasing the number of homogeneous exposure units allows an insurer to have greater confidence that its experience will not differ significantly from what is anticipated. Most importantly, this principle is essential in setting rates.[7]

The rate-making function, therefore, is inextricably tied to the ability of an insurer to predict, as accurately as possible, total losses, knowing that this prediction is subject to uncertainty. Herein lies one of the factors affecting insurance availability for lead hazard evaluation and control professionals.

Currently, limited actuarial data (information on lead related claims frequency and severity) exist for lead poisoning claims. In addition, with the legal, regulatory, legislative and medical landscapes still evolving, it is very hard to predict the results of future lead litigation.

Also, the state of the art for lead paint abatement and testing is still evolving, and the legal standard of care is following closely behind. There is concern on the part of insurers that improper testing and abatement practices will increase the hazards and result in unnecessary costs, including claims. Because the lead hazard evaluation and control field is a fledgling industry, insurers consider it to be high risk.

These and other fundamental issues, which go to the heart of the insurance mechanism, underscore why the aforementioned coverages are difficult to obtain.

HOW TO EVALUATE AVAILABLE COVERAGE

The vast majority of companies operating in the property and casualty insurance industry currently look at lead hazard contracting and consulting operations as commercially uninsurable. Fortunately, there are a small number of insurers offering this coverage. But even these firms are proceeding cautiously. A strong emphasis is being placed by the insurers on training, related experience, health, safety, and loss prevention in evaluating the lead hazard evaluation and control professionals who are venturing forth into this relatively new industry.

But these professionals must be careful to thoroughly evaluate the insurers offering the coverage as well as the policy forms available. Again, this insurance can only be obtained from a few insurers, and the coverage terms and pricing can vary greatly. Further, the financial and operational strength, including claims-paying ability, of the insurers must be taken into consideration when assessing the ability to respond to future losses.

Because insurance can be a complex subject for many individuals, the lead hazard evaluation and control professional should consult a knowledgeable insurance broker or even an attorney when evaluating lead-related contracting and consulting insurance coverages. The unusual nature of this insurance requires that the broker or attorney be familiar with so-called environmental or pollution liability coverages and the insurers that provide them. Together, the lead hazard evaluation and control professional and his or her insurance "expert" should ask the following questions:

What type of insuring "entity" is offering the coverage?

Is the insurer functioning as an admitted insurer, an approved but nonadmitted insurer, an offshore or alien insurer, or a risk retention group?

Because insurance is a highly regulated industry, the degree of regulatory oversight will depend on the organizational status of the insurer in each state. Generally, those insurers that are *admitted* in a particular state are much more closely regulated and are subject to participation in "guarantee funds," which are established to pay claims if an insurer should fail. *Approved but nonadmitted insurers* are not as closely regulated or may be minimally regulated, depending on the state involved. Further, with the exception of New Jersey, nonadmitted insurers are not subject to guarantee fund protection. The degree of regulation of an *offshore or alien insurer* will be dependent on the country of domicile; in fact, in some locations, regulation is almost nonexistent. *Risk retention groups* are permitted under federal law to transact insurance and are exempted from certain state insurance regulations to the extent that those regulations would otherwise make the groups unlawful. They are also not subject to guarantee fund protection.

How "solvent" is the insuring entity?

Solvency is the ability of an insurer to meet its current and future financial obligations by maintaining liquidity, adequate loss reserves, and appropriate premium rates.[8] The concept of solvency is particularly difficult to understand and, for that reason alone, the insurance "expert" should be relied upon to ferret out key financial data and indicators.

One source for this information is the statutory accounting report every insurer must file yearly with state insurance regulators. Additionally, there are several insurer rating and evaluation organizations, with the most frequently consulted being the A. M. Best Company of Oldwick, NJ. Best's service classifies insurers according to size, financial strength, underwriting results, and management strength. While Best's ratings are not always infallible, they can provide useful information about the insurer's performance.

What are the insuring entity's underwriting criteria, and how capable and knowledgeable are its staff and key managers?

Underwriting is the process of evaluating the potential for loss exposures, the selection of applicants based on risk characteristics and loss history, and the determination of an appropriate premium.

Generally speaking, those insurers having stringent underwriting guidelines are committed to controlling both short- and long-term loss potential. The significance here, in assessing the viability of an insurer, is to ascertain whether the insurer seeks only to cover those contractors and consultants that take steps to minimize loss through proper training and experience, and good safety and loss prevention techniques. Additionally, the knowledge of the insurer's staff about lead hazard evaluation and control methodology, regulations, equipment, and other issues translates into better, more responsive service and commitment to insureds.

What are the terms, conditions, definitions and exclusions in the insuring entity's policy?

Perhaps the most important exercise in this evaluation process is to read and thoroughly review the policy. This includes both the insurance "expert" and the lead hazard evaluation and control professional. Coverage can vary greatly from insurer to insurer and one must be careful to analyze the differences. Also, the policies must be written or endorsed so that no exclusions or other policy terms serve to preclude coverage for lead related exposures that are likely to arise from lead hazard control contracting, inspecting, risk assessing, designing, and planning activities.

The prospective insured must be comfortable with and aware of what the policy is and is not providing so that other informed risk management decisions can be made about potentially uninsured operations or exposures.

It is recommended that the insurers providing these coverages be approved as admitted or nonadmitted insurers by the state insurance department in the state where the lead hazard control work or contract is to be performed. Additionally, it is suggested that the insurers be rated as "A," Class VIII or higher by Best's.

SUMMARY

As a lead hazard evaluation and control industry continues to evolve, there will be a need for professionals entering and operating in that field to carefully assess and evaluate their potential liability and loss exposures. Therefore, a well-planned risk management program, to include several lead-related insurance coverages, is essential to properly addressing and treating those exposures.

Notes

1. Alliance to End Childhood Lead Poisoning. *A National Action Plan for Preventing Childhood Lead Poisoning.* Washington, DC, January 1993.
2. B. D. Smith, J. S. Trieschmann, and E. A. Wiening. *Property and Liability Insurance Principles.* Malvern, PA: Insurance Institute of America, 1987, p. 241.
3. Ibid., p. 251.
4. Ibid., p. 252.
5. From the 1986 version of the Insurance Services Office's Commercial General Liability Coverage Part.
6. Smith et al., *Property and Liability Insurance Principles,* p. 6.
7. Ibid.
8. T. E. Troxel and C. L. Breslin. *Property and Liability Insurance Accounting and Finance,* 2d ed. Malvern, PA: American Institute for Property and Liability Underwriters, 1983, p. 222.

16

Issues in Lead Poisoning Litigation

Christopher M. Placitella, Esq.

Barry R. Sugarman, Esq.

Litigation on behalf of lead poisoned children is fast becoming a major area in toxic tort litigation. Childhood lead poisoning is a problem of epidemic proportion in communities throughout the United States.[1] The federal government recently declared lead poisoning "the No. 1 environmental health hazard facing children."[2] More than 3 million children below the age of 6 have blood lead levels high enough to cause serious health problems.[3] Among large cities, New York City is estimated to have the highest childhood lead poisoning rate in the country.[4]

Children below age 6 are most susceptible to lead poisoning and its harmful effects. This is because their neurological and other body systems are in the early stages of development and absorb more lead from their environment than older children and adults. Young children are also more likely to contract lead poisoning because of their natural hand-to-mouth activity.[5] Lead poisoning, even at low exposure levels, causes permanent reductions in a child's intelligence, short-term memory loss, academic underachievement due to resulting behavioral problems and learning disabilities, hyperactivity, and brain dysfunction.

Contrary to widely held belief, lead poisoning is not a disease exclusively afflicting children of the poor. Wealth and status do not confer immunity from this insidious disease. Children from families above the poverty level comprise the largest group of persons at risk of exposure to this dangerous neurotoxin and its harmful effects.[6] Part of the reason lead poisoning disproportionately effects children from low-income families, however, is that these children's diets are often found to be nutritionally deficient.[7] A poorly nourished child can absorb as much as 50 percent of lead ingested, while a healthier, better nourished child will absorb only 5 percent of ingested lead.[8] Additionally, low-income families are more likely to live in older homes where deteriorating lead paint conditions cause their children to contract lead poisoning.

Several lead sources in the home present immediate dangers to young children. The primary source of lead in the home is lead-based paint.[9] Other sources include the home's water supply, soil, ceramic cookware, stove, older cribs and toys, and so on.

In 1978, the Consumer Product Safety Commission banned the manufacture and use of lead-based paint in residential housing. Unfortunately, this prohibition did not require removal of lead-based paint from surfaces previously coated with this dangerous neurotoxin. Consequently, today about 3 million tons of leaded paint remain in approximately 57 million occupied private housing units built prior to 1980. This represents approximately 74 percent of all such housing.[10] Despite legislative efforts to preclude future use of lead-based paint, the Agency for Toxic Substances and Disease Registry (ATSDR) states that corrective actions have been a complete failure.[11]

Lead from painted surfaces can enter a child's body completely undetected. Leaded paint naturally erodes to create a chalky dust and begins to chip and flake away from painted surfaces. The deteriorated leaded paint becomes part of the house dust and is inhaled by young children. The simple act of a child breathing in a leaded environment can begin the poisoning process.

The necessary corollary of the enormity of the childhood lead poisoning problem is that the victims will seek compensation from those believed responsible for their injuries. In the face of private and government inaction to remedy this pervasive problem, the legal system is being used as an instrument of protection and compensation for those stricken with this disease and related injuries. The primary defendants in lead poisoning personal injury actions include property owners (landlords), abatement contractors, and governmental entities.

What follows is designed to: (1) provide a basic examination of the theories of liability typically advanced in childhood lead poisoning litigation; (2) report on litigation activity in civil suits seeking monetary damages and injunctive relief believed to be representative of litigation activity occurring nationwide; and (3) identify some of the practical problems confronting the parties in this burgeoning area of litigation.

The reader must be aware that many jurisdictions have laws and regulations that can be effectively used in presenting childhood lead poisoning cases. Since these laws and their promulgated standards are not uniform, and in some states do not exist at all, they will only be addressed to the extent these apply to the reported cases or are helpful in advancing a particular theory of liability.

THEORIES OF LIABILITY

The primary causes of action advanced in lead poisoning litigation against property owners are found in negligence, strict liability, breach of contract, and violations of landlord-tenant law.[12]

The bases of the negligence claim are straightforward and really no mystery. The claim alleges that a defendant property owner (hereinafter "landlord"), or their agents, breached the duty to maintain the property where the plaintiff-tenant lived in a safe and sanitary condition. The duty is derived from several sources. These include the duty to act reasonably in light of the attendant facts and circumstances (the "reasonable man" standard), contractual obligation, statutes and regulations, principles of landlord-tenant law, and judicial precedent.

The most common basis for negligence claims stems from the plaintiff's allegation that the defendant landlord knew or should have known that conditions at their property, rented to the plaintiff's family, posed health risks to the plaintiff. Proof of notice, actual or constructive, is perhaps the most important part of the negligence claim.[13] There are several methods to substantiate a claim that a landlord knew, or that a reasonably prudent person in the landlord's position should have known, of the health hazards presented by their rental property. First, landlords frequently inspect the subject dwelling at the termination of a prior tenancy. Often the landlord, or their agent, conducts a "showing" or "walk-through" inspection of the dwelling at the outset of the plaintiff's tenancy. Many landlords collect their rents personally. In cases where the landlord receives the plaintiff tenant's rent from public assistance the payor agency may inspect the subject premises or require a statement from the property owner that the unit is in a safe and sanitary condition. Arguably, these acts place the landlord on actual notice of two important factors. These are: (1) the unit is in a deteriorated state that poses health risks to the future occupants, and (2) the unit will be occupied by a child. Establishing these two points is a crucial part of a plaintiff's liability case. Additionally, some jurisdictions hold landlords responsible for the conditions at their rental property, regardless of whether they actually knew of the hazardous conditions, as part of the responsibility to inspect and maintain the property in a habitable condition.

In *Acosta v. Irdank Realty Corp.*, 238 N.Y.S.2d 713 (1963), an early childhood lead poisoning decision, the court found the presence of these factors created notice of the hazardous conditions and held the defendant landlord was negligent and liable for the infant plaintiff's lead poisoning injuries. In *Acosta,* the court found that the cracking and peeling walls at the apartment owned by the defendant and rented to the plaintiff presented dangers to the plaintiff reasonably foreseeable to the landlord. Based on the combination of the broken walls and plaster, the "known propensities of children" to place everything in their mouths, and the presence of leaded paint, the court concluded the infant plaintiff's lead poisoning was the direct result of the defendant's negligent maintenance of the premises.[14] Significantly, the defendant was held liable for the plaintiff's lead related injuries notwithstanding the defendant's initial unawareness that the deteriorated paint was leaded. The court opined that once the landlord was aware of the dilapidated conditions at the property where the plaintiff resided, it was reasonably foreseeable that ingestion of painted plaster and peeling paint of any kind was potentially injurious.[15] Seven years later in *Garcia v. Freeland Realty, Inc.*, 314 N.Y.S.2d 215 (1970), the New York courts took judicial notice of the fact that in older apartments peeling and deteriorated paint is ingested by children, who then become lead poisoned and suffer severe medical consequences.

A landlord's notice of the hazardous conditions that caused the plaintiff's injuries remains the crucial proof in lead poisoning litigation. *Actual* notice of the condition claimed to have caused the plaintiff's injury may be required. This begs the question of what is meant by actual notice.

In *Brown v. Marathon Realty, Inc.*, 565 N.Y.S.2d 219 (A.D. 2 Dept. 1991), it was held the plaintiff tenant did not present facts sufficient to withstand the defendant's motion for summary judgment on the issue of actual notice of the conditions that allegedly caused the plaintiff's lead poisoning and resulting injuries.[16] The plaintiff presented evidence consisting of the landlord's deposition testimony he was inside the plaintiff's residence on numerous occasions while collecting rent and that the plaintiff informed the defendant of the results of a city inspection verifying the existence of lead-based paint hazards.[17] The appellate division reversed the trial court's motion determination and held that notice means more than actual or constructive knowledge. The court held that to impose liability upon a landowner for injuries resulting from a defective condition on leased premises the plaintiff must establish the landowner's notice of the condition for such a period of time that in the exercise of reasonable care he should have remedied it; or that the lessor retained sufficient control of the premises to impute constructive notice of the condition.[18] The *Brown* decision implies that proof of notice requires a showing of both knowledge (actual or constructive) of the condition and control of the effected premises. Parties defending future lead poisoning claims will surely seize upon the New York court's decision, along with any local landlord-tenant law and precedent, to combat lead poisoning claims predicated on negligent maintenance.

The landlord's duty owed to their tenant is also derived from statutes and local regulations. These provisions require state and local health and housing officials to inspect residences for the presence of lead-based paint where there is a report of a child with an elevated blood lead level, order abatement where appropriate, and compel uncooperative landlords to abate dwellings containing hazardous lead-based paint conditions. In some jurisdictions the agency responsible for the enforcement of the lead poisoning prevention laws is authorized to request legal action if the landlord fails to comply with an abatement order, and the court may impose substantial fines and penalties for the landlord's noncompliance.

The fundamental principle underlying these statutory and regulatory provisions is to clearly establish that the landlord bears the responsibility to make and maintain their tenant's dwelling reasonably safe condition for its intended purpose (i.e., residential housing). Plaintiffs often argue that the presence of and failure to abate lead conditions, in violation of these statutes and regulations, is negligence per se. It is likely, however, that statutory violations will be treated as evidence of negligence to be weighed by the trier of fact, which is most jurisdictions is the jury. The existence of these statutes further solidifies, however, the landlord's duty to protect the tenant from the risks of lead poisoning present at their rental property.

The aggressive enforcement of the statutory scheme for the prevention and control of childhood lead poisoning in some jurisdictions and the overall awareness and efforts of health and housing officials in jurisdictions where such statutes and regulations do not exist, combined with the tremendous publicity the lead poisoning issue has received in recent years, make it increasingly difficult for property owners to utilize lack of notice of the hazardous lead conditions as a defense. State and local health departments in several states are required by law to inspect residential housing for the presence of hazardous lead-based paint. Among these states are Massachusetts, New Jersey, New York, Maryland, Pennsylvania, Rhode Island, Ohio, and the District of Columbia.[19] If the inspection reveals lead-based paint conditions that pose health risks to the unit's occupants, the health department issues an abatement order to the landlord.

An abatement order issued to the landlord by a public agency may effectively negate a defendant's claim that they lacked the requisite notice or control over the property to fulfill the duties owed to a plaintiff tenant. This is precisely how a New York court ruled on the notice issue in a recently adjudicated lead poisoning case. In *Jessica Nunez v. KB Management, Inc.*,[20] the court entered an order granting summary judgment in favor of the infant plaintiff on the issue of liability where the medical records and health department inspection report conclusively proved the child was lead poisoned and his apartment contained lead-based paint conditions in violation of state and local law.[21] Once placed on notice, landlords face additional claims based on the child's continuing exposure and the landlord's legal obligation to abate the lead hazards within the time period prescribed in the order.

Negligence claims for a landlord's failure to abate or improper abatement of hazardous lead conditions are commonly included in civil actions filed on behalf of lead poisoned children. These claims include allegations that the landlord and his agents, usually a painter or contractor, failed to use reasonable care for the protection of the plaintiff when abating the plaintiff's residence and failed to observe the jurisdiction's regulatory guidelines for the safe removal of lead-based paint.

The negligent abatement claim is an integral component of lead poisoning personal injury actions because the lead abatement (repair and removal) process potentially creates greater dangers if it is performed improperly or if children are present before the complete removal of all traces of lead. The debris, loose paint chips, and dust generated by improper abatement contaminate the floors, rugs, and ventilation system. Children return to the home and are reexposed to this health hazard at levels exceeding the original exposure that triggered the abatement.

Many states maintain elaborate and stringent procedures for lead abatement, and federal guidelines exist that are used by many states lacking promulgated abatement standards. Regardless of the authority, however, most abatement standards have one goal in mind: to remove the lead from the affected area in a nonabrasive way that does not generate excess airborne dust or debris that could serve as a source of reexposure.

The negligent abatement claim is typically proven in one of three ways. First, a local health or housing authority's reinspection report indicates whether the landlord's abatement efforts were timely, complete, and performed properly. Health officials' notes and reports are often replete with references to incomplete or improper abatement activity. Second, plaintiffs often retain independent environmental experts to inspect and perform a risk assessment of a lead poisoned child's home after abatement is allegedly complete. Their observations, including pictures and paint sampling results, can be an extremely powerful piece of evidence against the landlord, or his agents, responsible for the abatement of a lead poisoned child's residence. The results of an independent risk assessment can also help create a cause of action for negligent inspection against the public entity charged with the responsibility of inspecting the plaintiff's residence if it is found they incorrectly certified the subject premises as properly abated. Lastly, by chronologically tracking the plaintiff's postabatement rising blood lead levels, an indication of more recent exposure, it becomes clear that the abatement was performed improperly and that the plaintiff has been further injured by reexposure.

A breach of contract claim is usually included in lead poisoning complaints. This claim alleges that the defendant landlord violated the express terms of the lease by failing to keep the plaintiff's residence in a safe and sanitary condition and free from the hazardous lead conditions. The breach of contract count can be the easiest to prove by referral to the terms and covenents in the lease between the defendant-landlord and the plaintiff-tenant's family. Many standard residential leases expressly require the lessor (landlord) to properly maintain the plaintiff's residence. Alternatively, leases not containing express provisions of this kind often contain terms prohibiting the tenant from independently making certain types of repairs or decorations. Among these are structural alterations and decorating projects such as painting and hanging wallpaper. These contractual provisions imply that the landlord has retained the exclusive right and responsibility to keep the subject premises' structure in good repair to the exclusion of the plaintiff-tenant.

COMMON PROBLEMS IN CHILDHOOD LEAD POISONING LITIGATION

In addition to the myriad of generic factual and legal difficulties encountered in all personal injury litigation (i.e., theories of causation, damages, experts testimony, etc.), lead poisoning cases present particular problems for their litigants. Two of these are (1) the counterclaim or third-party complaint against the infant plaintiff's parents and (2) the great amount of documentary evidence against and "slumlord" label often attached to landlord defendants.

The negligent supervision claim typically alleges the plaintiff's parent(s) failed to prevent their child from ingesting leaded paint and that this failure was the proximate cause of the child's injuries. This can be an effective defense tool—to place an obstacle between the plaintiff and his guardian, the parent. If plaintiff's counsel does not act quickly to dismiss the claim, they may be burdened with a conflict of interest requiring the appointment of a guardian for the infant plaintiff for purposes of the lawsuit and the parent(s) retention of separate counsel to defend the counterclaim/third-party complaint. Since the majority of plaintiffs are members of lower income households, this places strain on their already limited resources and can be an effective deterrent to continuing the lawsuit.

Plaintiff's motion to dismiss or application for summary judgment is predicated on at least three factors. First, that the defendant does not possess any facts indicating a lack of parental supervision. Second, the plaintiff must inform the court of the *fact* that ingestion of lead paint does not necessarily mean the infant plaintiff ate lead paint chips. The court must be made aware of the voluminous medical and legislative findings that lead poisoning most often results from normal hand-to-mouth activity and inhalation of leaded dust from deteriorated leaded surfaces. The fact that a child was exposed to dangerous lead conditions in the home and contracted lead poisoning as a result in no way implies parental misconduct. Finally, the plaintiff must argue that parental immunity shields the plaintiff's parent(s) from the allegations raised by the counterclaim/third-party complaint. Many states have case law shielding parents from liability for negligent acts within the scope of customary child care absent a showing of gross negligence or willful and wanton misconduct.[22] These decisions are premised on the notion that the preservation of the family unit demands that activities involving parental discipline, care, and control must remain free from judicial intrusion.[23]

Landlords encounter the dilemma of refuting the significant amounts of unfavorable and objective evidence that is almost always present at the initiation of the lawsuit and dealing with the "slumlord" label and its accompanying stigma.

The evidence includes health department inspection reports indicating the existence of lead-based paint hazards in violation of both state and local law, inspector's notes of correspondence and conversations with the defendant regarding abatement, the defendant's own conduct towards the plaintiff's family (e.g., retaliatory eviction attempts), an independent risk assessment, medical reports diagnosing lead poisoning and that the plaintiff's current problems are lead related, and so on. These baseline proofs are usually present prior to commencement of formal discovery.

The portrayal of the defendant landlord as a slumlord, which in some cases may be appropriate, is a significant problem for the landlord during discovery and before a jury at the time of trial. This is particularly true in multiple dwelling cases or in cases where the landlord owns significant amounts of nonowner occupied rental property. By proving the defendant's conduct in the case at bar and the benefits the defendant derives from property ownership, the plaintiff accentuates the relative inequality of the parties. The inference created is that the landlord's acts in the particular case, and overall course of conduct, took advantage of his superior position and caused the plaintiff's injuries. The opportunity to present the aforementioned objective evidence and portray the defendant as the villain is often readily available.

ACTION FOR INJUNCTIVE RELIEF

The class action suit for injunctive relief is being used in several jurisdictions in an effort to compel public entities to properly enforce their particular lead poisoning statutes and regulations.[24] One such action that is pending is *LaToya White, et al. v. City of Newark, et al.* (hereinafter "*White*").

The *White* class action seeks mandatory injunctive relief against the City of Newark, to compel enforcement of state and local lead poisoning laws regarding the City of Newark's alleged: lack of mandatory lead screening of all children ages 1 through 5; failure to inspect residential housing for the presence of lead-based paint; failure to require owners of lead contaminated premises to abate the hazardous lead conditions; and failure to ensure that abatement has been properly effectuated. The class consists of all children under age 6 who either live or lived in housing in the City of Newark built before 1978 that contained lead-based paint products.

Newark's Childhood Lead Poisoning Problem

According to recent estimates based on information compiled by the U.S. Public Health Service, the City of Newark, NJ, has the second highest lead poisoning rate among large cities in the country. Newark's lead poisoning rate is second only to New York City.[25]

The City of Newark is designated a Priority I cleanup community under state and federal standards due to the stock of older, deteriorating housing and the presence of demographic factors that indicate danger of lead poisoning.[26] All children who live in a Priority I community are by definition "at high risk of lead poisoning" and the city is required to screen these children for lead poisoning at least once a year. The majority of Newark's residential housing is deteriorating and more than two-thirds of the occupied housing was built prior to 1940.

Newark's Failure to Enforce Lead Poisoning Laws

The representative plaintiffs in the *White* action, then 3-year-old LaToya White and 4-year-old Teaire Adams, suffer from permanent neurological injuries due to their exposure to deteriorated lead paint in their Newark apartments. It is alleged that their injuries, as well as the injuries of numerous class members, are in part a direct result of the City of Newark's failure to adequately enforce state law, state department of health regulations and Newark city ordinances designed to control and prevent lead poisoning.

The plaintiff class alleges that the city, its agencies and certain officials have failed to: maintain a comprehensive lead poisoning screening program; conduct adequate and timely inspections of homes of children with elevated blood levels; comply with statutory and regulatory provisions concerning the declaration of a public nuisance where lead paint conditions are found in the dwelling of a lead-poisoned child; properly perform its duty to order landlords to complete lead abatement in a safe and timely fashion; complete abatement of dwellings where persons of high risk of lead poisoning are found and the landlord refuses or fails to complete abatement; monitor and supervise lead abatement procedures; and provide ongoing medical management of children and education and counseling of parents of lead poisoned children. The city is required to perform each of the above functions for the benefit of its residents pursuant to state and local law and regulation.[27]

Discovery

The parties are continuing to take extensive discovery including, but not limited to, substantial document production requests, written interrogatories, and depositions. The plaintiffs plan to produce testimony from the numerous class members and their families, members of the medical community who routinely interact with the Newark's Lead Poisoning Prevention Program, and submit documents obtained from the defendant and the New Jersey State Department of Health to illustrate the city's failure to comply with the state's laws and regulations. Ultimately, the plaintiffs hope the court will recognize the city's noncompliance and both order the city to immediately begin completely and properly fulfilling its public duties and devise a monitoring program to ensure such compliance.

The City of Newark's Position

The city contends they have properly and adequately enforced and complied with the state and local laws, regulations, and ordinances relevant to the action. The city, in its pretrial memorandum, asserts that its screening program fulfills all statutory and regulatory requirements, it conducts all inspections of homes within one week of receiving a report of a lead poisoned child, and that it is properly monitoring all abatement activities and reinspecting affected dwellings for lead removal prior to the landlord repainting the abated surfaces.

Plaintiffs' Current Findings

The primary information the parties believe will substantiate their respective positions is the medical records and inspection reports contained within the files of the city's Lead Poisoning Control Program. The city recently produced these documents that outline a substantial portion of its lead poisoning prevention record on a case-by-case basis. The files consist of more than 15,000 pages of blood lead level results, inspectors' notes, abatement orders, requests for legal action, and so on, and the plaintiffs' representatives are reviewing these documents. The preliminary review of a limited portion of these records indicates, however, that the city has failed to properly and timely inspect and order abated numerous properties where lead poisoned children and children defined as "high risk of lead poisoning" reside.

Documents received from the state department of health reveal serious deficiencies in the functioning of the Newark's Lead Poisoning Prevention Program. Reports from site inspections conducted by state health officials show that the city has screened less than 20 percent of the children requiring lead screening and that the city has failed to coordinate its efforts with private physicians and that city lead inspectors have falsified records. State health officials also observed problems with environmental investigations and unacceptable protocol for follow-up of these investigations.

RECENT VERDICTS AND SETTLEMENTS

Lead poisoning causes a variety of serious injuries. These include learning disabilities, behavioral problems, hyperactivity, attention disorders, cognitive deficits, and, in the most severe cases, organic brain damage or death. These injuries affect a child's academic performance, social progress, and future employability. The damages alleged include claims for physical injury, pain and suffering, emotional distress, medical surveillance, and future lost wages. The future lost wages claim is necessarily a large part of these damages claims due to the plaintiff's young age and long life expectancy. The reported New York verdicts and settlements reflect these special factors. A 1989 plaintiff's verdict for $1.7 million was recently found not to be excessive and upheld by the appellate division.[28] There are five reported case settlements for figures ranging from $300,000 to $912,500.[29]. In April 1992, a defense verdict was returned in a case where the prevailing defendant admitted the plaintiff was lead poisoned but alternatively claimed the condition was congenital and that the source of the lead was not the defendant's building.[30] In May 1993, after summary judgment was granted for the plaintiffs on all liability issues, a case was settled for $250,000.[31] Similar verdicts and settlements are being reached throughout the United States. A Connecticut landlord was recently found liable for damages due to his negligent failure to keep the premises rented to the plaintiff's family free of lead paint where it was found that he should have known lead paint was present and that a child resided at the premises. The jury awarded the plaintiff almost $930,000.[32] An Ohio appellate court recently affirmed a lower court ruling in favor of a defendant landlord where the family of a lead poisoned child could not establish the landlord's actual or constructive notice of the presence of lead paint in the apartment.

LEAD PAINT INDUSTRY LIABILITY LITIGATION

As outlined above, landlords, abatement contractors, and public entities are currently the primary defendants in childhood lead poisoning litigation. A more expansive view of the childhood lead poisoning problem and the resulting litigation, however, requires examination of its root causes, the production and distribution of hazardous lead-based paint for use in residential housing. Litigation in process throughout the country finds injured parties attempting to recover for their injuries resulting from exposure to lead-based paint manufactured and distributed by the lead pigment companies, and their modern day corporate successors, responsible for producing, marketing, and distributing lead-based paint for use in residential housing.[33]

The plaintiffs in these actions allege that the defendant paint companies and lead pigment producers were aware of the unreasonable risks to young children posed by the manufacture and distribution of lead-based paint for use in residential housing as early as the 1920s, but ignored these dangers and made a decision in furtherance of corporate profit to continue manufacturing, marketing, and distributing lead-based paint. The claims against these corporate defendants are predicated upon both traditional and less traditional theories of liability including negligent product design, negligent failure to warn, breach of warranty, and concert of action.

The negligent product design claim alleges that the lead pigment manufacturers produced and continued to produce lead-based paint for use on interior and exterior surfaces of residential housing notwithstanding their knowledge of the unreasonable health risks and hazards that exposure to these products presented to young children. Documents discovered in this litigation indicate this decision was based solely on profit and completely disregarded the known health risks attendant to the customary use of its products.

The negligent failure to warn claim alleges that the lead pigment producers had a duty to warn the consuming public to whom its products were marketed—including suppliers, manufacturers, retailers, and parents of young children—of the foreseeable risks created by exposure to their products, but failed to do so. Industry documents indicate medical personnel working with the Lead Industries Association (LIA) advised them of the necessity to institute a program to warn parents about the dangers associated with lead poisoning and stated that no new mother should be permitted to leave the hospital without being handed information regarding the risks of infant lead exposure. The plaintiffs allege that no such action was taken and that the price of the lead industry's inaction is the poisoning, which was entirely preventable, of an entire generation of children.

The breach of warranty claim alleges that the lead pigment producers breached the express and implied warranties that their lead-based paint products were safe and fit for their intended purpose, covering interior and exterior surfaces in residential housing, and that they did not act to prevent the foreseeable dangers to those who could reasonably be expected to use and be exposed to these products.

The concert of action theory of liability claims that the lead pigment producers and paint companies all belonged to a trade association known as the LIA, which conspired on behalf of all of its members to conceal and mislead the general public regarding the risks posed to young children by the lead products manufactured, refined, marketed, and distributed by its members. While medical personnel and officials of the association and its members were acutely aware of the grave health risks associated with infant lead exposure, it is alleged that they not only failed to remedy this problem but continued to market and promote the leaded products known to be its cause.

The greatest obstacle to suits against lead paint companies and lead pigment producers to date has been the plaintiffs' inability to identify with certainty the producer, manufacturer, or distributor of the lead paint product that is claimed to have caused the plaintiff's lead poisoning and related injuries. While the body of medical and scientific knowledge has evolved to the point that a child's injuries can be causally related to their lead exposure, determining which company's product caused the plaintiff's injuries remains problematic. For this reason, plaintiffs are urging courts to hold all members of the relatively limited number of lead pigment producers jointly and severally liable for their injuries under the theory of market share liability.

The market share theory as applied in lead litigation argues that approximately six lead pigment producers marketed and sold all, or virtually all, of the lead used in lead-based paints sold in the United States.[34] It further argues that since plaintiffs are unable to identify which lead pigment producers caused their injuries due to factors completely beyond their control, each lead pigment producer should be held liable and their liability should be apportioned according to each defendant's market share of the lead pigment produced. Those factors include: (1) the application of several successive layers of lead paint to surfaces in the homes where the plaintiffs reside; (2) the lack of information as to the precise formulas or other identifying characteristics of the various lead pigments produced by the lead pigment manufacturers; and (3) the fact that the lead-based paints are substantially similar in composition.

To date, courts have declined to adopt the market share theory to apportion liability among lead pigment producers.

The product identification riddle "Whose lead was it?" may, however, soon be unraveled. Lead is an element with both common and distinguishing characteristics. Lead ore mined in different areas of the country can be chemically dissected for the presence of additional trace metals and analyzed by examination and comparison of ratios of its stable isotopes. Some researchers in the geochemical community believe that this type of analysis of a particular lead paint sample may reveal a "lead fingerprint" that can be traced to a particular lead pigment manufacturer's product. This method of lead source identification has not been attempted in litigation but is expected to be advanced by those who would hold the industry liable for its apparent negligent and improper conduct in the production and marketing of lead-based paint.

CONCLUSION

The legal system is beginning to react to society's recognition that childhood lead poisoning is a problem of epidemic proportion, stunting the growth of an entire generation. As increasing numbers of children are diagnosed with lead poisoning, it is inevitable that lead poisoning cases will occupy a significant part of the litigation landscape. While many theories of recovery exist, it is clear that, for the moment, traditional negligence theories remain most viable. Plaintiffs and defendants alike face serious obstacles to the successful presentation and defense of their cases. The complexion of these cases remains unclear, particularly as it relates to suits against the lead paint industry, but it is virtually certain they will be vigorously litigated until the legal community reaches consensus on their viability.

Notes

1. The authors represent clients in lead poisoning litigation in New Jersey and New York but are aware of significant amounts of lead poisoning litigation activity in Massachusetts, Maryland, Washington, DC, Louisiana, Michigan, Pennsylvania, Rhode Island, Illinois, and Wisconsin.
2. *The New York Times,* February 21, 1991, quoting (then) Health and Human Services Secretary Louis W. Sullivan as the Bush Administration announced a $1 billion five-year plan to reduce toxic lead exposure in children.
3. Ibid.
4. "Legacy of Lead: America's Continuing Epidemic of Childhood Lead Poisoning," Environmental Defense Fund, March 1990.
5. "Preventing Lead Poisoning in Young Children—A Statement by the Centers for Disease Control," October 1991.
6. Agency for Toxic Substances and Disease Registry (ATSDR). *The Nature and Extent of Lead Poisoning in Children in the United States: A Report to Congress.* (July 1988).
7. "When Will We Stop Poisoning Our Children," *Boston Globe Magazine,* February 21, 1988 (quoting Dr. John Graef).
8. Ibid.
9. Ibid at p. 18–20. Lead-based paint is not the only source of this hazardous element

found in and around the home. Lead from a dwelling's water source (i.e., well or supply facility) or from deteriorating lead pipes or copper pipes with lead solder is also a significant cause of childhood lead poisoning.

10. U.S. Department of Housing and Urban Development (HUD). *Comprehensive and Workable Plan for the Abatement of Lead-Based Paint in Privately Owned Housing: Report to Congress.* Washington, DC, 1990.
11. ATSDR, *Nature and Extent of Lead Poisoning.*
12. There are several additional causes of action including, but not limited to, gross negligence, nuisance (strict liability), consumer fraud, ultrahazardous activity (lead abatement). These are called to the reader's attention but are beyond the scope of this article.
13. This sets aside the difficult damages and causation issues associated with the injuries alleged in lead poisoning cases in favor of a discussion of liability factors.
14. 238 N.Y.S.2d at 715.
15. 238 N.Y.S.2d at 714.
16. Ibid at 220–221. The Appellate Division reversed the Supreme Court's denial of the defendant's motion for summary judgment and dismissed the plaintiff's complaint.
17. Ibid.
18. See generally, *Putnam v. Stout,* 38 N.Y.2d 607 (1976); *Franqui v. City of New York,* 152 A.D. 2d 482 (1978); and *Tobar v. City of New York,* 146 A.D.2d 694 (2d Dep't 1989).
19. This list is not exhaustive of those states with legislation or pending legislation and the reader is urged to consult local authorities.
20. NY Sup. Ct., Bronx County, Case No. 7931/91.
21. *Jessica Nunez v. KB Management,* NY Sup. Ct., Bronx County, Case No. 7931/91.
22. *Franklin v. Krumanocker,* 494 N.Y.S.2d 214 (A.D. 3 Dept. 1985); see also *Holodook v. Spencer,* 36 N.Y.2d 35 (1974); see also *Foldi v. Jeffries,* 93 N.J. 533 (1983) and *Mancinelli v. Crosby,* 247 N.J. Super. 456 (App. Div. 1991) (citing *Foldi* with approval for the principle that reasons exist for the retention of the parental immunity doctrine in areas involving parental authority and customary child care.) See *Goller v. White* 20 Wis.2d 402, 122 N.W.2d 193 (1963).
23. *Foldi v. Jeffries,* 93 N.J. 533, 545 (1983).
24. See also *Coalition to End Lead Poisoning v. Koch,* 138 Misc.2d 188, 524 N.Y.S.2d 314 (1987) and see also *Darrick & Deborah Wade, et al. v. Cuyahoga Metropolitan Housing Authority, et al.,* Case No. 1:92CV1596, (N.D. Ohio, E. Div.)
25. "Legacy of Lead." Estimates were of children 6 months to 5 years of age with blood lead levels of over 10 μg/dl.
26. The prioritization of New Jersey communities was accomplished in 1987 by utilizing data from the National Health and Nutrition Examination Survey (NHANES II) and a Community Potential Hazard score that incorporated specific socioeconomic and demographic variables associated with childhood lead poisoning.
27. New Jersey's Lead Poisoning Prevention Statutes and Regulations are codified at *N.J.S.A.* 24:14A–1, et seq., *N.J.S.A.* 26:2–130, et seq., *N.J.A.C.* 8:51–1.1, et seq., *N.J.A.C.* 8:52–5.2. The references to local ordinances are to Newark Revised Ordinances 14:2A–2, et seq., 15:2–1, et seq., and 15:9–2, et seq.

28. *Miller v. Beaugrand-Fisher Group, Inc., et al.,* 564 N.Y.S.2d 390 (1991); appeal denied 571 N.Y.S.2d 913.
29. See Jury Verdict Research.
30. *Hill, Pro Ami v. City of New York, et al.,* NY Sup. Ct, New York County, Case No. 10414/89 (April 1992).
31. *Jessica Nunez v. KB Management,* NY Sup. Ct., Bronx County, Case No. 7931/91 (May 1993).
32. *Hardy v. Griffin,* 569 A.2d 49 (CT 1989).
33. *Santiago v. Sherwin Williams Co. et al.,* Case No. 92-2263 (D. Mass. 1987).
34. These companies, and their predecessors in interest, are Sherwin-Williams Company, NL Industries, Inc., Eagle Picher Industries, Inc., SCM Corporation, Atlantic Richfield Company, and the Lead Industries Association.

17

Selecting a Lead-based Paint Abatement Consultant

Vincent M. Coluccio, Dr. P.H.

Adapted from the National Lead Abatement Council's Technical Report Number 2, June 1992, by Marc P. Vink and Maureen A. Sullivan

State and local agencies are increasingly requiring contractors working in the lead-based paint (LBP) abatement field to be trained, certified, and licensed. While Massachusetts and Maryland led the way with legislation that addresses LBP topics, more needs to be done. Further, significant differences exist in LBP testing, abatement, final clearance, and worker protection between the public housing, the private sector, and the commercial area. Understanding present regulations, keeping abreast of changing state and local requirements, and discerning the critical differences in various technical considerations are essential for any consultant whose operation covers the broad spectrum of the emerging lead abatement industry. Hopefully, as states join to enact LBP legislation, they will have identified these significant discrepancies and adopt comprehensive LBP laws. Finally, LBP abatement practices and procedures have not been fully developed and codified. In some instances, poor abatement design, unacceptable field procedures, lack of familiarity with federal and state regulations, and inadequate investigation have lead to unnecessary costs and created a greater degree of contamination than previously existed. These situations have resulted in damage to property, bodily injury to workers, facility occupants, community residents, and litigation.

The purpose of this chapter is to familiarize facility owners, property managers, and other interested parties with the fundamental questions that should be addressed in identifying experienced and qualified LBP consultants.

PURPOSE OF TECHNICAL REPORT

Facility owners, property managers, and other interested parties must properly prepare to select an LBP consultant. It is crucial to define what the LBP consultant is to accomplish. In some cases, this may be very general, but in other cases, it can be a very detailed design program. Only after the facility owner and/or property manager have clearly defined the role(s) the LBP consultant is to fulfill will he/she be equipped to judge which consultant is best qualified to serve the project needs.

It may be useful to compile readily available information about your facility, including all potential sources of LBP contamination and any records that may help to define where LBP may have been used. A little time spent in preparing this information can improve your understanding of your facility or property needs and could also lead to enhancement of your consultant's comprehension of the project.

It is often desirable to establish procedures and policies that will be universally applied for the selection of an LBP consultant. Although selection criteria may change from project to project, the selection process for a particular project should be clearly established in advance and made available to all interested parties.

UTILIZING AN LBP CONSULTANT

The LBP consultant should be able to guide the facility owner and/or property manger through a variety of important professional issues including: (1) LBP identification, (2) assessment, (3) regulatory compliance, (4) properly conceived and documented response action, (5) options, and (6) possible selection of an LBP analytical laboratory, abatement contractor, waste transporter, and appropriate waste disposal facility.

With the exception of step 6, these five steps ultimately lead to the development of a specification and drawing that acts as an "instruction" and/or "blueprint" for the project. It is very important that project specifications and drawings be explicit and detailed as they will act to define the scope of work performed by the contractor and abatement support team.[1]

Additionally, adequate on-site contractor project management and quality assurance is essential for a smooth and efficient operation.[2] The LBP consultant must be able to interpret and manage abatement specifications, solve a variety of problems that arise, and be able to assist the contractor and support team in completing the project efficiently.

DEFINING A PROFESSIONAL

Merriam Webster's Collegiate Dictionary defines a *professional* as:

> 1*a*: of, relating to, or characteristic of a profession *b*: engaged in one of the more learned professions *c*: characterized by or conforming to the technical or ethical

standards of a professional 2*a*: participating for gain or livelihood in an activity or field of endeavor often engaged in by amateurs *b*: having a particular profession as a permanent career *c*: engaged in by persons receiving financial return 3: following a line of conduct as though it were a profession.

It is clear from *Webster's* definition that a professional is an individual who possesses special knowledge and skill resulting from a combination of education and experience in a particular branch of science or learning.[3] It is also important to recognize that a professional has the legal obligation to perform the duties of his or her profession in accordance with the standards of that profession as measured by the knowledge, skill, and performance of his or her peers.

CONSIDERATIONS FOR SELECTING LBP CONSULTANTS

Lead-based paint abatement projects demand a high degree of special knowledge and experience from all members of the abatement support team, including the consultant, laboratory, abatement contractor, waste transporter, and waste disposal facility. The LBP consultant must often direct many members of the abatement support team to ensure successful completion and fulfillment of contractual obligations.

The most important element in successfully completing an LBP abatement project is the selection of the best available individual or company to engineer and perform the work. If a firm is selected, make sure that the project manager is of the highest quality.[4] Even though the firm may have successfully demonstrated its capability, it is important that qualified and experienced staff are assigned to your project. While no single method of selecting an LBP consultant is foolproof, there are several fundamental methods that can be employed to determine an LBP consultant's knowledge and experience.

Education and Certification

The consensus of certifying bodies, such as the American Board of Industrial Hygiene (ABIH) and the Board of Certified Safety Professionals (BCSP), is that an individual must have at least a baccalaureate degree in chemical or physical science. In lieu of this, a similar discipline with five years of full-time employment in the professional world is desirable to qualify for sitting for a designation examination. The American Academy of Environmental Engineers (AAEE) requires a formal education, state issued professional engineer's license, and eight years of environmental engineering experience, coupled with an examination, interview, and peer review process for specialty certification.[5]

Although it may be difficult to find an LBP consultant who has all these qualifications, the closer to this definition one gets, the more one can be sure that the individual satisfies the consensus standards established by ABIH, BCSP, and/or AAEE for professionalism. Always ask for a copy of the consultant's resume. An LBP consultant interested in employment will supply you with information about his or her qualifications. This document affords you the opportunity to verify both education and experience. Additionally, you should request that a copy of the consultant's degree, designations, and any other support materials accompany the resume. Be sure to verify degrees, designations and other support materials for authenticity.

LBP Training

In addition to establishing such fundamentals as baseline education, designations, and experience, it is important to identify any continuing and/or specialized training the individual has relevant to LBP abatement. There are as many as 50 or more LBP training centers located throughout the United States. Training centers are represented by private commercial enterprises that range in varying degrees of accountability, with a main focus toward profit, to 14 university programs with funding from the National Institute for Occupational Safety and Health (NIOSH) and the Environmental Protection Agency (EPA), with the main focus being education.

At present, most available training programs range from six-hour worker training to five-day contractor training. The programs that follow the EPA model curriculum are clearly desirable. The EPA model accreditation program for inspectors and supervisors has recently become available, and these basic programs should be considered mandatory criteria in your LBP consultant selection process for residential housing abatement. In addition, the Steel Structure Painting Council (4400 Fifth Ave., Pittsburgh, PA) has developed and sponsors an LBP abatement training program especially designed for the steel structures painting industry.

Project Experience

A successful history of completed LBP abatement projects is a compelling indicator of experience. This is particularly true if the dimensions of prior projects correspond to the project being contemplated. However, if the dimensions do not correspond to the project being contemplated, previous project experience may not be as meaningful. For example, successful completion of residential projects does not demonstrate commercial and/or industrial capabilities. Conversely, successful outdoor work does not demonstrate indoor experience.

There are many consultants who have excellent credentials and experience with one or more environmental problems. These same people may have little experience with LBP. It is both frustrating and expensive to discover this after the fact. Since there is no magic formula for assuring that the consultant you hire will be a perfect match for your project, it is vital that you do your homework.

Most consultants will supply you with a list of their most successfully completed projects, but such a list may not be truly indicative of the quality of work performed. Ask the consultant for a complete job list and follow up, on a random basis, to carefully examine references. Careful examination of five or more unsolicited and randomly selected references may be the best predictor of future job performance. Checking references for the firm as well as for individuals assigned to your project is important. It is also important to pay attention to any subcontractors and vendors the consultant may propose to hire. Form 17-1 is provided as a sample consultant's reference check form that may prove helpful in accumulating and comparing reference check information.

Insurance

At a minimum, LBP consultants should carry workers' compensation, general liability, and motor vehicle insurance. By virtue of the fact there are only a few guidance documents, few established competency criteria, and a body of emerging legislation relative to LBP abatement, the potential for legal entanglement is great. Because of this, it is highly desirable that LBP consultants have professional liability insurance covering errors and omission. While this type of insurance is generally available for most LBP consultants, it is not readily available from domestic A-rated insurance companies.

Further, some professional liability policies may not provide coverage for so-called pollution or contamination claims. A-rated insurance companies, representing a financially stable segment of the insurance market, may utilize rigorous underwriting practices to identify insurable risks.

The LBP consultant with a history of claims, poor credit, and limited experience may find it difficult to purchase the most desirable type of professional liability insurance and may have to settle for less desirable coverage from off-shore carriers, risk retention groups, risk purchasing groups, or self-funded programs. An insurance professional should be consulted to help appraise the quality of a consultant's insurance program.

Professional Affiliations

Membership in a professional society and/or trade association may not in itself be reflective of an LBP consultant's competency. This is particularly true if the affiliation is passive. However, if an LBP consultant serves on a professional society and/or trade association committee, regularly publishes articles, or speaks professionally, it could be an indicator of his or her dedication to, and involvement in, the discipline. Although there are numerous professional societies and/or trade asso-

FORM 17-1. Consultant's Reference Check Form

Consultant's Name ______________________

Reference Name/Title/Company ______________________

Type of Work ______________________ **When Performed** ______________________

1. Relationship to consultant

( ) a. employee/relative
( ) b. client
( ) c. responsible for hiring
( ) d. consultant/other _____

2. Experience Level of Reference Contact

( ) a. none
( ) b. minimal
( ) c. by association
( ) d. degree/designation level______________ M.S./Ph.D./Arch./P.E./CIH/CSP

3. No. of yrs. of Association

( ) a. none
( ) b. < or = 1
( ) c. > 2 or < or = 3
( ) d. > 3

4. Key items used in selecting this consultant

( ) a. price /bid
( ) b. references
( ) c. prequalifications
( ) d. experience
( ) e. other

Value of work performed:

Discuss the consultant's performance in the following categories, and determine if the reference believes his/her performance was: poor (P), good (G), excellent (E), or does not apply (DNA).

	P	G	E	DNA
Project Management/Control	( )	( )	( )	( )
Work Procedures and Practices	( )	( )	( )	( )
Knowledge of Regs/Specs	( )	( )	( )	( )
Documentation/Reports	( )	( )	( )	( )
Credibility (Honest, Trustworthy, etc.)	( )	( )	( )	( )
Quality of Management	( )	( )	( )	( )
Quality of Staff	( )	( )	( )	( )
Communications/Cooperation	( )	( )	( )	( )
Overall Quality	( )	( )	( )	( )

On a Scale of 1–10, 10 being highest, how would you rate the consultant. _________

Questions to be used during the conversation:

Would you hire/recommend the consultant again? ______________________

In what area does the consultant excel? ______________________

In what area can the consultant improve? ______________________

Additional Comments: ______________________

ciations that have an interest in LBP issues, the most notable are: the National Lead Abatement Council, the American Industrial Hygiene Association, the Steel Structures Painting Council, the National Association of Professional Painters and Decorators, the Environmental Information Association, and the Alliance to End Childhood Lead Poisoning. These organizations sponsor trade shows and special symposia on LBP issues as well as publish periodicals for their members.

Cost

Consultants typically use two methods of charging for their work. The method of payment you select may depend on the type and scope of the project.

The most common type of contract consultants use is the *cost plus fixed fee contract.* This method charges for salaries and expenses plus a fixed fee to be determined by you and the consultant. This method may work well for projects where the scope of work is unclear, extensive investigation is needed, or experimental processes are used.

The other type is the *fixed price method.* This method specifies a firm sum for the entire project. The method is generally used for small projects or when the scope is clearly defined.[6] Whatever method is agreed upon, be sure to establish clear, specific criteria to evaluate and compare cost estimates.

Assembling an LBP abatement team can be very expensive, particularly if approached from an "on-the-job training" perspective. Since a poorly designed and executed project can create greater contamination and be a greater health threat than no abatement at all, price should be a secondary consideration. There are many consultants who will engage an LBP abatement project without having the proper education and training, project experience, and insurance. This lack of quality may very well be reflected in a low bid and all of the possible consequences that could accompany it.

Documentation

A completed project file is one of the most significant features used to determine the professional level and quality of performance by a consultant. How complete are these files? Do they contain information on initial lead inspections, chain of custody, air monitoring reports, wipe sample reports, notification letters, project specifications, project minutes, test results, waste manifests, daily checklists, and fire and safety meetings? Are the records available, completely organized and consistent, as they relate to regulated versus nonregulated jobs, large versus small projects?

Comprehensive documentation can be an effective first-glance appraisal of the matter of work habits. Orderly record keeping is a positive characteristic of an organized firm.

The assertion that a consultant is responsible for his or her work and workers may be substantiated in the project files. Conversely, unorganized records, or documentation that is haphazard and incomplete, will convey a negative message to any potential client. If you are paying for the services of a professional, you should demand a professional product.

In the event of any possible future litigation, it is imperative that complete documentation be available for all work completed by the consultant, since such records would serve as the primary element in the defense against claims. Ideally, these records should be categorized, dated, signed, labeled, and fastened in a binder. Furthermore, these records should be stored in a secured, fireproof area.

Contracts

An LBP consultant should provide you with an estimate of anticipated project costs and an itemized fee schedule. "Iron clad," signed contracts that define services and fees are important elements in establishing the business relationship.[7]

Failure to adhere to mutually agreeable contracts leads to adverse relationships. Request contracts written in simple English. They should also clarify penalty and interest fee schedules. An attorney specializing in contract law should be consulted to help you draft contract language.

CONCLUSIONS

It would be fair to say that, at present, the LBP abatement market has not yet been fully besieged with substantial opportunities in an emerging industry. Indeed, even where abatement is both necessary and mandated, economic shortfalls prevent some areas of the industry from moving ahead. Sizeable contractors with prior abatement experience and smaller companies that may be new to the industry are engaging the services of a consultant to train and assist them on a day-to-day basis while actively involved in an abatement job. This growing reliance on a consultant's competence dictates that they be well versed in the latest technology and laws.

The lack of regulations, accredited teaching facilities, and project experience could hamper the LBP abatement market before it has chance to achieve its potential. If the present trend continues, consultants could play a critical role. Not only could they act as the initial source of education for many contractors in the LBP field, but their advice could help to define a feasible, strategic approach to lead abatement.

Although there are several nationally recognized authorities and many more qualified individuals on the subject of LBP abatement, there are certainly more unqualified individuals. When selecting an LBP consultant or contractor, watch out for hard-sell approaches, conflicts of interest, overly optimistic timeliness, and strong biases for or against certain abatement practices, and cost estimates *significantly* lower than those of competing firms. Do not take anything for granted.

The cost of LBP abatement is a significant consideration. However, even though cost is important, it should never overshadow the knowledge, experience, and professionalism of LBP abatement or general remediation contractors or consultants. In considering that a poorly performed LBP abatement project can create more contamination and be a greater threat to public health than no abatement at all, price becomes even less of a consideration. There are many contractors and consultants—without the proper knowledge, experience, and professionalism—who will engage a LBP abatement project. Their lack of quality may be reflected in a low bid, and all of the consequences that accompany it. Check out the individual or company thoroughly, and then check again.

Notes

1. W. Nylin and W. Tansil. Choosing a consultant. *Asbestos Issues*, December, p. 32, 34 (1988).
2. J. Krukowski. Calling in the cavalry. *Pollution Eng.* February 15, p.43 (1992).
3. D. Malecki, J. Donaldson, R. Horm, *Commercial Liability Risk Management and Insurance*, Vol. II. American Institute For Property and Liability Underwriters, December 1981, p. 157.
4. K. Roberts. Bypassing the consultant can hurt a project, *ECON*, August 1991, p. 14, 39.
5. *Environmental Engineering Selection Guide 1991*. Annapolis, MD: American Academy of Environmental Engineers, p. 11.
6. *Selecting an Environmental Consulting Firm.* Portland, OR: Department of Environmental Quality, p. 2.
7. J. D. Roberts. So you would like to become a consultant? Consider the costs. *Pollution Eng.* December 1990, p. 36.

18

Selecting a Lead-based Paint Abatement Contractor

Vincent M. Coluccio, Dr. P.H.

Adapted from the National Lead Abatement Council's Technical Report Number 1, August 1991, by Marc P. Vink and Maureen A. Sullivan

According to a December 7, 1990, report to Congress, the Department of Housing and Urban Development (HUD) estimates that there are approximately 57 million privately owned and occupied homes built before 1980 that contain lead-based paint (LBP). Of these units, approximately 9.9 million are occupied by families that have children under the age of 7. Additionally, 3.8 million of these units have peeling paint, excessive amounts of dust containing lead, or both problems. As previously thought, older homes are more likely to have LBP than new homes. An estimated 90 percent of dwellings built before 1940 have LBP in the interior or on the exterior, while 62 percent of homes built between 1960 and 1979 have LBP.

Lead is a toxicant with no known purpose in the human body. Although lead can injure virtually all parts of the body, the primary target organ is the central nervous system. Blood lead levels of 80 micrograms (μg) per deciliter (dl) can be manifest as kidney failure, gastrointestinal problems, coma, convulsions, seizures, and pronounced mental retardation in children. Blood lead levels of 40 μg/dl or higher can manifest as reduced hemoglobin and mild anemia. Blood lead levels from 10–15 μg/dl and 25 μg/dl may manifest as slower reaction time, reductions in intelligence and short term memory, neurobehavioral deficits, and adverse effects on heme biosynthesis and on vitamin D and calcium metabolism.

Most states and local governments have done little to respond to LBP hazards, except to react to individual cases of childhood poisoning.

In many instances, the improper abatement of LBP has created a greater contamination than existed before abatement took place. At present, there are very few contracting firms nationally that have successfully completed a LBP abatement project.

IMPORTANT QUESTIONS TO BE ASKED

Given the extent of LBP in housing, the toxic effects of lead, and the dearth of regulations and abatement contractors, the selection of experienced and/or qualified LBP abatement contractors is an important public health issue. While no method of selecting an LBP abatement contractor is foolproof, there are several fundamental questions that should be asked of any person or company performing specialized work.

How long has the company been involved in general environmental remediation and/or specifically in LBP abatement?

Presently, there are relatively few companies that have performed LBP abatement; therefore, building owners may be forced to seek services from general environmental remediation firms or other building or interior renovation contractors.

In order to accommodate the emerging LBP abatement market, some general environmental remediation firms are developing their staffs to perform this type of work, while others have hired highly experienced personnel. Conversely, there are also many general environmental and remediation firms that are poorly staffed and have never performed a LBP abatement project but are willing to try.

What projects has the company completed that are of similar dimensions to the project contemplated?

Often, companies will supply interested parties with a list of successfully completed projects. These projects may act as a testament to the quality of work performed.

A careful examination of references from previous projects may be the single best predicator of future job performance. If not already provided, ask for: (1) the project location, (2) date(s) of project start and completion, (3) description of the project including scope of work, and (4) name and telephone number of building owner and/or representative in charge of the project. This information could act as the foundation for additional data gathering and will provide a historical perspective of the contractor's performance. Form 18-1 is an example of an organized form that may prove helpful in the accumulation and comparison of reference check information.

While most firms will supply you with a list of their most successfully completed projects, such a list may not be truly indicative of the quality of work performed. Careful examination of references from previous projects becomes even more important when they are unsolicited and randomly selected. Ask the contrac-

FORM 18-1. Contractor Reference Check

Contractor Name ________________________________

Reference Name/Title ________________________________

Value of Reference (Check Answer Which Most Closely Matches the Response of the Reference)

1. Relationship to Contractor

( ) a. Employee/Relative
( ) b. Supplier
( ) c. Industrial Hygienist (IH)/Consultant
( ) d. Hired the Contractor(s)

2. Abatement Experience

( ) a. None
( ) b. Minimal
( ) c. Building Inspector, Management Planner, Worker/Supervisor training
( ) d. Industrial Hygienist, Engineer

3. Number of Jobs with Contractor

( ) a. None
( ) b. 1 or 1 month
( ) c. 2–10 or 1 year
( ) d. > 1 year or > 10 jobs

4. Key Items in Selecting a Contractor

( ) a. Price/Bid
( ) b. Equipment
( ) c. References
( ) d. Prequalifications

Discuss the contractors performance in the following categories, and determine if the reference believes their performance was poor, good, excellent, or did not apply.

5. Value of Work Performed	Poor	Good	Excellent	DNA
Timeliness or Scheduling	( )	( )	( )	( )
Abatement of Work Procedures	( )	( )	( )	( )
Knowledge of Regs/Specs	( )	( )	( )	( )
Recordkeeping/Documentation	( )	( )	( )	( )
Housekeeping (cleanliness, organization)	( )	( )	( )	( )
Quality of Supervisors	( )	( )	( )	( )
Quality of Workers	( )	( )	( )	( )
Communications/Cooperation	( )	( )	( )	( )
Overall Quality	( )	( )	( )	( )

Questions to be used during conversation:

Would you hire contractor again? ________________

In what areas does the contractor excel? ________________

What areas can the contractor improve? ________________

Additional Comments:

____

____

____

tor for a complete job list for a given period of time, then follow up on random selections from this list.

What is the general environmental remediation and/or specific LBP abatement experience of key on-site personnel?

The quality of work performed is highly dependent upon the experience level of such key people as site superintendents, project managers, supervisors, and other hands-on personnel. Even if the president of the environmental remediation firm has ample experience, this will not reflect in quality of work performed if he or she is not at the job site in a hands-on-capacity.

What are the LBP abatement education and training requirements for company personnel?

An LBP abatement contractor wishing to perform work for HUD must submit proof of training by an institution providing at least three days of instruction. Similarly, LBP abatement contractors performing work in Maryland and Massachusetts must submit proof of minimal training. However, be cautioned that what is acceptable to HUD or in Massachusetts or Maryland may not necessarily be acceptable for some abatement projects or acceptable in other states.

Depending upon the dimensions of the project, building owners may require all or only some workers or key personnel trained in LBP abatement. Request a copy of the training certificate that is often issued with completion of the course, but be aware that such a certificate may not be an endorsement of proficiency by the institution issuing the certificate.

Does the environmental contractor have well developed safety and health programs?

Such programs should include respiratory protection, medical surveillance, and specific LBP standard operating procedures. Additionally, does the contractor have quality control (QC) and quality assurance (QA) programs that may indicate the firm's capabilities and performance?

Insist on copies of the above programs and read through them to get a better appreciation of the depth of planning that has gone into these documents. If it appears that these documents have been drafted hurriedly or without much forethought, it may indicate that the firm's dedication to safety and health, QC, and QA may not be all that has been represented.

Does the environmental contractor have a good environmental and/or occupation safety and health record?

A history of significant Occupational Safety and Health Administration (OSHA), Environmental Protection Agency (EPA), and other regulatory agency citations and/or violations may indicate poor management infrastructure. In some cases, citations and/or violations may be minor and, as such, do not necessarily indicate poor management infrastructure. However, if citations and/or violations are issued frequently and are represented by large fines, request additional information and be very cautious.

Environmental and occupation safety and health citations and/or violations histories are available through the Freedom of Information Act from a variety of governmental agencies; Appendix C, "Sources of Information," contains listings of federal and state regulatory agencies that may prove helpful in examining a contractor's regulatory compliance history. This information may also be available from other state or local agencies such as local health departments.

Does the environmental contractor have Commercial General Liability (CGL) insurance that covers the company's LBP abatement operations?

Beyond the required Workers' Compensation Insurance, contractors should have CGL. Standard CGL insurance policies typically contain absolute pollution exclusions. Unless the environmental contractor has special endorsement stating that LBP abatement is covered in the policy, it is probably not covered. In consideration of the fact that insurance definitions, exclusions, and endorsements are composed of specific language, obtain a copy of the policy for examination by your insurance specialist, attorney, or other qualified professional.

Is the insurance company offering CGL coverage for LBP abatement financially stable and well managed?

Since LBP insurance is not available through the standard insurance market, other insurance vehicles such as risk retention groups, risk purchasing groups, and offshore nondomestic markets have emerged. Although these alternative vehicles may be appropriate, check the company's financial standing and reinsurance. This is most easily accomplished by asking for the insurer's annual statutory financial statement or "convention blank." This booklet provides financial information such as loss reserves, reinsurance recoverables, investment portfolio, annual written premium, and admitted assets.

Does the contracting company have good financial credit?

Depending upon the size of the project being undertaken, bad credit history, especially as it relates to equipment and material supplies, could delay a project from completion in a timely manner. Failure to obtain credit may impair a company from supplying all the necessary equipment materials and labor.

If the project being undertaken is a large one, it may prove helpful to call the contractor's equipment and material supplier(s) to determine credit worthiness. Additionally, you may wish to check with your local Better Business Bureau.

Does the contracting company have sufficient bonding capacity?

Depending upon the size of the project, a contracting company may need bonding. Verify the contracting firm's bonding capacity. Make sure the bonding company or surety is domestic and U.S. Department of the Treasury T-Listed. Determine if additional bonding is available to the contracting company in case an unforeseen increase in the scope of the project develops. Additionally, you may also wish to determine if there are any instances of revoked bonds.

Depending on the size of the project and the availability of time, interview the contractor at the firm's place of business. Ask specific questions concerning your project. You may also want to meet with several of the contractor's managers to ascertain their philosophy of doing business and to develop a sense of their knowledge, experience, and professionalism.

While at the contractor's place of business, look at the firm's administrative and warehouse facilities. Check to see if good housekeeping (cleanliness, organization) practices are maintained and that equipment and supplies are well cared for. Does the operation appear to be well managed, first-class, or is it a fly-by-night operation?

If at all possible, visit at least two or three on-going jobs. Observe the quality of the site supervisor, worker behavior, and general work practices and procedures.

Selected Reading

Buford, John L. Selecting your asbestos abatement contractor. *Professional Safety* 15–17 (Feb. 1991).

Cowan, Zachary S., III. Basic considerations for selecting asbestos abatement contractors. *Natl. Asbestos Council J.* 39–41 (spring 1990).

Gray, Julian C. Lab selection—More than a game of chance. *Asbestos Issues* 32–36 (Aug. 1989).

Nylin, Bill, and Wayne Tansil. Choosing a consultant. *Asbestos Issues* 32–34 (Dec. 1988).

Olander, Robert C. Good Help Is Hard to Find. *Asbestos Abatement* 63–64 (Jan./Feb. 1988).

19

Interim Control of Lead-based Paint Hazards

David Chambers

Title X of the Housing and Community Development Act of 1992 defines interim controls as:

> a set of measures designed to reduce temporarily human exposure or likely exposure to lead-based paint hazards, including specialized cleaning, repairs, maintenance, painting, temporary containment, ongoing monitoring of lead-based paint hazards or potential hazards and the establishment and operation of management and resident education programs.[1]

If properly monitored and frequently evaluated, interim controls may be an effective step in controlling lead-based paint hazards in buildings. However, the presence of lead-based paint, no matter the condition, poses a potential threat to human health and therefore interim controls must be accompanied by a plan for eventual full-scale abatement.

The use of trained (and eventually accredited and certified) professionals is essential for establishing a program that both protects occupants from lead exposure and reduces a building owner's liability. Once an interim control process is implemented, each phase of the process should be examined closely by a knowledgeable individual to assure the effectiveness of the process or to recommend necessary corrections. Property owners have a duty to protect occupants from lead-based paint hazards, and, when full scale abatement is not feasible, interim controls are the first step in an overall program of lead hazard management.

This information was obtained from ATC Environmental Inc.'s direct experience in managing lead hazards and from the ongoing national process of developing new federal guidelines for in-place management of lead-based paint.

Regardless of the condition of lead-based paint, abatement is the preferable treatment. However, economics mandate that other measures be taken to control the immediate or potential hazard. Those dwelling units that have no noticeable substrate faults and wherein the current application of lead-based paint is in good condition may be the best candidates for interim control measures. The building's maintenance practices and procedures are also a major consideration when determining what method of lead-based paint hazard control should be used. If preventive maintenance is practiced on a frequent timetable, the appropriateness of an interim control program increases, and the building owner's costs for the program decreases.

On the other hand, those units that are structurally unsound or damaged may need immediate abatement of lead-based paint. Deterioration of substrate and paint coatings due to water damage is a common problem in many residential units, and interim control measures will not be effective in those instances until the source of the water damage is repaired. In such cases, abatement may have to take place to correct damage, thus negating the need for other controls on the affected substrates. An evaluation by a trained and experienced risk assessor is even more valuable in these situations. The risk assessor may determine that basic window repair may not adequately control the lead hazards and may recommend total replacement of the window as the only safe and effective way to handle the hazard posed by lead paint.

Interim controls are only effective when the scope of the lead paint problem is known. An inspection for lead paint is not only strongly recommended, but in the near future may be required by regulatory authorities before interim control measures can be taken. If a building owner chooses to forego a full inspection, then duty-of-care liability forces the owner to assume that all surfaces contain lead-based paint. For example, the City of New York assumes that all units built prior to 1960, where a child under the age of seven resides, contains lead-based paint. A building owner who adopts this type of position may be in for a costly, and perhaps partially unnecessary, lead management program. A thorough inspection and risk assessment can identify those building components that must be managed through the use of interim controls and those that must be abated.

RECOGNIZING LEAD-BASED PAINT PROBLEMS

Almost all lead-based paint used is oil-based. The primary cause of damage to lead-based paint is water, either through the obvious pipe leak or the less obvious condensation of the surface or substrate. Since oil and water do not mix, a wet substrate will not accept the paint, creating a gap between the two surfaces. The normal heating and cooling cycles found in every building create humidity, which pushes the paint away, and drying, which causes the paint to contract. This constant movement promotes paint failure, leading to blisters, cracks, and, eventually, loss of sections of paint. Once an opening in the paint is created, more moisture can soak into the substrate and enhance the loss of integrity.

The normal aging process is also a culprit in paint damage. After a certain period of time, the flexibility of paint fails, creating a resistance to the normal expansion and contraction of the substrate. This resistance leads to cracking that may, in turn, cause a release of lead dust and chips. Even paint in a well-maintained unit is subject to damage through aging. This unavoidable process reinforces one of the most important aspects of the interim control program: periodic surveillance.

Painting history in a single unit will vary widely from other units in the building. This history may be crucial in the overall ability of the paint to remain intact and in good condition. Latex paint applied over an oil-based paint will eventually fail, although it may be some time before this failure is noticed. Surface preparation is also key to the ability of paint to adhere to a substrate. Grease or dirt on a surface may repel the paint, causing eventual failure. One badly prepared layer of paint may eventually cause all subsequent layers to crack or peel away. For all of these reasons, periodic maintenance and observation will ensure that potential hazards are dealt with immediately. A paint in good condition today may not be in a week.

Physical damage to paint is often the most obvious problem in a dwelling. It may be caused by a door that swings open, unhindered by a door stopper. A child's play room may also evidence contact damage that can be easily spotted and thus quickly repaired. However, damage through friction may be less noticeable but no less threatening. Cabinets, drawers, windows, and doors may all have surfaces that continually grind against each another, releasing fine particles containing lead into the environment. This occurrence is especially a problem in older dwellings where the number of paint layers is high. A cabinet door that once swung freely may now have so many layers of paint that it will not close without a hard push. Each push may release lead dust. Multiply the number of pushes in a year, and the need for some control measure becomes apparent.

Finally, accessibility of the painted surface to a child must be addressed. Although the paint is intact and in seemingly good condition, one of the primary advantages of lead-based paint is its ability to "chalk." A microscopic outer layer of the paint sloughs off, creating a new, fresh-looking surface. This action may be further enhanced by a child's hand or body. One interim solution may be to repaint with a nonleaded paint, but prior to taking this action, the surface should be examined closely for bite marks. If they are present, immediate abatement should take place because a layer of paint will not prevent or reduce the hazard. Any protruding surface should be thoroughly checked before deciding on interim control measures.

INTERIM CONTROLS FOR DAMAGED PAINT

The first step in controlling undesirable existing paint conditions is to find the source of the problem and repair it immediately. If there is an indication of water damage, the sources of that damage may be many. Frequently the cause may be a bathroom or kitchen above the affected dwelling. Recaulking tubs, sealing cracks in the floor, or replacing missing tiles and grout may solve the problem. In addition, educating tenants to report these problems promptly and taking immediate action to repair the damage may save a building owner hundreds or even thousands of dollars.

Other interior water problems may result from leaking refrigerators, clogged drains, lack of drip pans under water heaters, and air conditioning units. Rooftop air conditioning units should be frequently cleaned and maintained by the installation company. Window units are a common cause of water damage and should be positioned so that condensation falls away from the building. It is also a good idea to seal around the unit with insulation to prevent rain from entering the dwelling.

Many dwellings have terraces or balconies that may or may not have floor drains. If there is a floor drain, it, along with any gutters or piping used to draw the water away from the building, should be cleaned regularly. For terraces without a drain, there is often a gentle slope to allow the water to fall away from the building. Where artificial grass or carpeting on terraces traps water, such coverings should be removed.

A complete examination of the roof is necessary because roof leaks are another common source for water damage. Particular attention should be given to the gutters and downspouts. These may need to be replaced or simply cleaned out. The roof flashing should be scrutinized for holes or weak spots, especially in corners or along the edges. Caulking of skylights or vents may be necessary, and frequent examination of any repairs will be an integral part of the interim control program.

Repairing the damaged paint to an intact condition is a relatively simple process, but if care is not taken to control dust and debris, a potential hazard could become an immediate one. Even for as simple a procedure as wet scraping a small area of peeling paint, a building owner should use personnel trained in lead-based paint procedures. Lead-based paint should never be removed without proper precautions (see Chapter 10). Cleaning the abated surface prior to repainting is critical to help remove the fine layer of lead dust caused by normal chalking activities as well as any gross debris left by the abatement activity. The use of a HEPA-filtered vacuum and wet cleaning methods should accomplish the task. Prior to repainting, the abated area should be sealed with a joint compound or spot sealant, whatever is appropriate to the task.

Painting the affected area should not be done until it is dry and free of any residue. An appropriate primer should be applied to the entire surface to ensure adherence of the new layer of paint. It is not advisable to paint only the affected area. The entire room or dwelling should be painted to control future exposure. Thus, the removal and patching of damaged paint should be viewed as not only an immediate hazard reduction, but a general interim control for the entire space.

The same methods used for cleaning and priming damaged paint should be used when a building owner chooses repainting as an interim control of surfaces that are undamaged. Exposed lead-based paint, regardless of its condition, always poses a potential for lead exposure.

The choice of paint used as the top coat will depend on a variety of factors, including cost. Unfortunately, choosing a cheaper paint may, in the end, cost far more since it will have to be patched and repaired frequently. Marine paints (mercury-free and lead-free) offer extremely durable finishes that are long-lasting. However, they may contain other toxins that would be inappropriate in residential use. High gloss paints offer the next level of protection and are much easier to wipe clean during the periodic maintenance procedures. A latex flat finish is the weakest solution and, in some cases, may actually be wiped off during routine cleaning.

While a well-prepared substrate and the use of a quality primer will definitely extend the life of the paint, the need for continual monitoring and spot maintenance will not be negated.

Protruding surfaces, such as baseboards, chair rails, and door frames are frequently removed as an interim control measure. This approach is usually the quickest and easiest way to deal with the components. However, costs can escalate when a large number of dwellings is involved. The risk assessor should determine the extent of damage and examine the components for bite marks. If the components are in good condition with no indication of mouthing activity, cleaning and repainting may be an appropriate choice.

If the extent of damage is severe or there is an indication of mouthing activity, removal or covering is the most appropriate response. Generally speaking, covering the component(s) is as expensive as removal and may actually be more labor intensive. If removal is chosen as the response, then appropriate abatement techniques must be followed since dust and debris generation may be high.

INTERIM CONTROLS FOR FRICTION SURFACES

Friction surfaces are those surfaces subject to rubbing or grinding from another surface. The most common friction surface component in a dwelling unit is a window, which generally has the highest levels of lead-based paint. Other friction surface components include doors, cabinets, stair railings, and floors. The continuous rubbing of these components may generate high levels of lead dust and constitute a major portion of a building owner's interim control program.

A lead-painted window is often the number one source of lead dust in a home.

This is especially true in older buildings where the many layers of lead paint are allowed to accumulate in the window well or sill. Every time a window is opened, the stop bead and parting bead are rubbed, releasing lead dust. The wood in older windows is often cracked, providing an excellent repository for lead dust.

Removing lead paint on windows is difficult, expensive, time-consuming, and may cost more than actual replacement. In older windows, the removal of lead paint may actually destroy the integrity of the wood. In that case, replacement should be considered. Because of the potential danger posed by lead-painted windows, a building owner must assess the options available and choose the one that offers the greatest gain in the long run.

Selection of options is often dictated by occupant patterns. Those tenants with infants or toddlers may receive replacement windows first, since their children are at the peak hand-to-mouth period and may be more susceptible to lead poisoning. In other units, abatement of friction surfaces and removal or covering of window sills may be adequate to protect the children. Working closely with the risk assessor is crucial in making these decisions.

Interim control measures on windows will almost always include some form of abatement. Therefore, appropriate preparation and tenant protection is mandatory to accomplish the task. Any abatement should be conducted by individuals trained in lead paint abatement methodologies. Key elements of interim control abatement on windows include:

- Wetting of all window components
- Wetting and removal of stop bead and parting bead
- HEPA vacuuming and wet removal of all loose and peeling paint
- Repairing the weight and pulley system to ensure a smoother opening and closing
- Removal of sashes
- Cleaning the window well or covering it with aluminum coil stock, caulking and sealing
- If possible, installing replacement window channels
- Repriming and repainting other surfaces

In certain cases a building owner may choose to cover friction surfaces with aluminum or other impenetrable covering materials, thus combining abatement and interim controls. The interior window sill should either be replaced, covered, or stripped as thoroughly as possible and then repainted following procedures discussed earlier in this chapter. A building owner must work closely with the risk assessor to determine the appropriate response to lead-painted windows since they may be the major source of lead contamination in the dwelling.

Friction damage to doors is most often caused by either too many layers of paint or a door frame that is no longer in alignment with the door itself. In some cases, damage is due to both reasons and must be addressed accordingly. A simple visual assessment will usually pinpoint the problem.

For doors that are out of alignment with the frame, simply rehanging the door may solve the problem. If multiple layers of paint is also a cause of friction damage, the door may need to be planed in order to fit properly within the frame. Planing will generate high levels of lead chips and dust and should be done away from the dwelling or in a fully protected area. Once the removal is complete, the door should be primed and repainted following all necessary safeguards. Obviously, replacement of the door is a permanent solution, as long as the door jamb is unaffected by the new door.

Drawers and cabinet doors are another common cause of friction-related lead dust in a home. The simplest solution is to remove the component and have it stripped off-site, primed, and repainted prior to installation. Examine the cabinet itself for friction damage and plane or cover as needed. Installation of bumpers will greatly reduce friction damage.

Painted floors, stair risers or treads, and decks may be large contributors to lead contamination. The most obvious solution is carpeting, but, if improperly installed, it may rub against the painted surface creating additional friction. Also, carpeting is extremely difficult to clean effectively, so a building owner may want to consider polyurethane or tile to cover the painted surface. In the case of painted stairways, be sure to cover not only the tread, but the riser as well since the toe of a shoe can create significant impact damage.

INTERIM CONTROLS FOR LEAD IN DUST

Lead in dust has been recognized as a major source of lead exposure in children. Lead in dust can come from any number of sources, even in dwellings where a visual assessment determines that all lead-painted surfaces are intact. Once cleaned, dust levels can rise quickly because of open windows, pets tracking in exterior dust, or countless other methods. We are surrounded by lead and eventually some of that lead will find its way into our homes. Interim control of lead dust is an integral part of a properly managed lead hazard reduction program.

Tenant education is key to controlling lead dust since no landlord can be expected to HEPA vacuum and mop a dwelling on a regular basis. Tenants should be instructed about the hazards of lead dust and informed of what they can do to reduce potential exposure for their children, such as frequent cleaning of upholstered and nonupholstered furniture, regularly scheduled wet-mopping with detergent, storage of toys and other belongings off the floor, changing of air conditioner filters, and scheduled washing of draperies and curtains.

However, tenant cleaning may be inadequate to control lead dust hazards. In that case, periodic professional lead dust cleaning should become a part of a building owner's interim control program. This professional cleaning, coupled with routine maintenance by the tenant, offers the best possible control of lead dust available. The landlord should notify tenants prior to professional cleaning and recommend that the tenant vacate the dwelling during cleaning activities. In addition, the tenant should be informed that the dwelling must be as clean as possible and free of objects on the floors to allow a professional contractor to accomplish the most thorough cleaning job possible. Due to the sensitive nature of such an operation, open discussion between the landlord and tenants is crucial to successful lead dust control.

The first step in controlling lead dust is to identify possible dust traps. These include window sills and wells, carpets, HVAC filters, upholstered furniture, radiators, grates, and cracks and crevices. Exterior building components and play areas where children have direct contact with soil or play sand are also areas where buildup of lead dust may be problematic. A risk assessment that includes dust wipe sampling is highly recommended before undertaking an interim dust control program.

Cleaning of hard surfaces such as tile or wood floors is usually accomplished by a HEPA vacuum and wet cleaning using trisodium phosphate (TSP). Note that TSP is illegal in some localities and if that is the case in your community, a lead-specific cleaning agent or strong detergent may be substituted.

There are a few key points to note when cleaning lead dust:

- Start at the highest point in a room and work down to the floor. The floor should be the last horizontal surface cleaned in the room.
- It is strongly recommended that two buckets be used for wet mopping, one to contain cleaning solution and the other to hold clean rinse water. The rinse water should be changed frequently, at least once per room.
- Take care that the exhaust unit of the vacuum does not stir up lead dust on the floor. This agitation can be controlled by laying down a sheet of polyethylene under the unit.
- Have enough disposable sponges or cloths so that they may be changed frequently during cleaning. Never reuse these.
- For smaller areas, pour the cleaning solution directly onto the sponge or cloth to prevent contamination of the cleaning solution.
- Always rinse the area after cleaning with a clean sponge, cloth, or mop.

Carpeting can be a major trap for lead dust and may be impossible to clean completely. If at all possible, all carpeting should be removed and disposed of in accordance with applicable rules and regulations. A building owner may discover that the attempt to clean carpeting of lead dust is as, or more, expensive than replacement. Should the attempt be made, there are a few recommendations to follow:

- HEPA vacuum with an attachment that rakes the carpet and loosens as much lead dust as possible. Vacuuming should be done very slowly.
- Steam clean using a phosphate-based detergent, if local regulations permit. Professional steam cleaning is recommended since that equipment is usually more powerful than typical rental equipment.

Upholstered furniture shares the same characteristics as carpeting; it may be impossible to clean thoroughly. If some pieces are known to be contaminated with high levels of lead, it may be best to dispose of the piece rather than attempt cleaning. Meticulous HEPA vacuuming should be performed and if possible, the piece covered in a washable slipcover. Although none of these actions will prevent lead exposure, they may be useful in lowering the potential for lead contamination to children who frequently use the furniture.

Lead dust in air handling systems may be a potential hazard. Filters should be replaced frequently and register vents cleaned with a damp sponge. For vent registers sealed to the wall by accumulated layers of paint, particular care should be taken if these registers are removed for cleaning. The act of removal may dislodge paint around the register. Exposed duct work can be easily cleaned using a HEPA vacuum. If cleaning within the duct work is required, it should be accomplished by a professional who understands and utilizes proper lead abatement and containment techniques.

The initial removal of lead dust from a dwelling is a vital step in lead poisoning prevention. Given the tendency for lead dust to reaccumulate on the very surfaces that have been cleaned, a planned maintenance and cleaning program is the only way for a building owner and a tenant to ensure that all possible precautions are being taken to prevent lead exposure in children. The cleanup of lead dust is one more element in an effective interim control program, and a schedule of future cleanings must be determined by the building owner, tenants, and a trained risk assessor.

ENCAPSULATION AS AN INTERIM CONTROL

The use of encapsulation as an interim control is fast becoming a popular method. Encapsulation should not be confused with enclosure, which is a rigid barrier mechanically fastened to the substrate. The use of encapsulation as a form of interim control has recently been approved by the state of Massachusetts.[2]

Spray-applied or painted-on encapsulants have long been a favorite of the asbestos abatement industry and may, if used appropriately, serve as an effective interim control for lead-based paint. Encapsulants in the lead industry often employ some sort of reinforcement material (i.e., fiberglass mesh). Success depends upon an assessment of the compatibility of the painted surface to accept an encapsulant, the likelihood of a bond between the encapsulant and the painted surface, and the susceptibility to contact or friction damage.

Encapsulant technology can be categorized as (1) nonreinforced liquid coatings applied by a sprayer or roller, (2) reinforced liquid coating (using a cloth mesh or fibers), and (3) materials applied directly to the painted surface with an adhesive (floor tile). The choice of method will depend on the compatibility of the encapsulant and painted surface, expected or anticipated life of the encapsulant, ease and safety in application, and compliance with environmental and building code requirements.

There are a number of factors to consider when choosing an encapsulant. For substrates that expand and contract, such as wood, the encapsulant will need to be flexible. Surfaces exposed to frequent contact or friction may require a stronger encapsulating agent. The level of skill required to apply the material may be a consideration, as well as the encapsulant's ability to be repaired easily by maintenance personnel in the building. Finally, the appearance of the encapsulant, while not a major factor in determining the use of an encapsulant, must be taken into consideration.

The application of encapsulants may require specific surface preparations. The product manufacturer will most likely recommend test patches be done to determine correct application, curing time, adhesion, and cohesion. In most applications, surface preparation is much like that done prior to painting: cleaning and "roughing up" of the surface, removal of loose paint, and possibly the use of some sort of primer.

In spite of some manufacturers' claims, encapsulants are still considered an interim control in most states. Like all other methods of interims controls, encapsulants must be closely monitored and maintained in order to act effectively as a barrier between lead-based paint and occupants. Only limited information is currently available on the long-term durability of encapsulants.

EDUCATION AND PERIODIC SURVEILLANCE

The use of interim controls will only be effective if accompanied by a building- or complex-wide tenant education program. Allowing tenants to take part in the hazard control process can be a cost-efficient and politically savvy way for a landlord to deal with the lead-based paint problem. Implementing this program is not a simple task, but it is hoped the advantages will outweigh the difficulties in the long run. The educational program can be divided into two different areas: notification and tenant responsibilities.

Notification may take many forms. Tenants should be notified well in advance of a risk assessor's visit to allow time to prepare for the visit and afford the landlord the opportunity to explain what will take place during the risk assessment. Questions or problems can be discussed prior to the risk assessment and issues brought up in the discussion may be useful in determining what interim controls or abatement methods are most appropriate and most convenient.

Upon the discovery of lead-based paint hazards, tenants should be notified so that they may take necessary actions. For example, once lead is discovered in the soil, tenants can instruct children not to play in affected areas. If lead is discovered in the dwelling, the tenant has the opportunity to control the child's behavior that could likely result in lead exposure. Finally, if an abatement is to occur, the occupants must be given ample time to have their belongings covered or moved out of the affected area.

A building owner should reach out to tenant organizations that may be quite effective in educating building occupants. The Environmental Protection Agency and most states have numerous pamphlets that will answer most concerns a parent or tenant may have regarding lead poisoning. Learning the actions to be taken required to maintain a lead-free or lead-controlled environment is the first step in prevention of childhood lead poisoning. It is up to the building owner and the tenant to take advantage of all educational programs available.

Once interim controls are in place, periodic surveillance is the next step in the lead management program. The schedule for periodic surveillance is usually determined by the risk assessor and carried out by the building maintenance personnel. In certain cases, the building owner may request that the risk assessor conduct periodic surveillance and train the maintenance personnel in what to look for in the way of potential problems.

Notes

1. Title X: Residential Lead-Based Paint Hazard Reduction Act of 1992, Section 1004 (13).
2. The Act Further Regulating Lead Paint in the Commonwealth (H 5302), signed by Massachusetts Governor William F. Weld on January 14, 1994.

Appendixes

Appendix A

Acronyms

AAEE	American Academy of Environmental Engineers
AAS	Atomic absorption spectroscopy
ABIA	American Board of Industrial Hygiene
AIHA	American Industrial Hygiene Association
ALC	Apparent lead concentration during XRF testing
A2LA	American Association for Laboratory Accreditation
ASTM	American Society for Testing and Materials
ASV	Anodic stripping voltametry
ATSDR	Federal Agency for Toxic Substances and Disease Registry
BAL	British Anti-Lewisite
BCSP	Board of Certified Safety Professionals
CAA	Clean Air Act
CDC	Federal Centers for Disease Control and Prevention
CERCLA	Comprehensive Environmental Response, Compensation, and Liability Act
CFR	Code of Federal Regulations
CLC	Corrected lead concentration during XRF testing
CPCU	Chartered property casualty underwriter
CPSC	Federal Consumer Product Safety Commission
DHHS	Department of Health and Human Services
DQO	Data quality objectives
EBL	Elevated blood lead
ELAP	Environmental Lead Proficiency Analytical Testing Program
EP	Erythrocyte protoporphyrin
EPA	Federal Environmental Protection Agency
FHA	Federal Highway Administration

HEPA	High efficiency particle air (filter)
HHS	Federal Department of Health and Human Services
HUD	Federal Department of Housing and Urban Development
IARC	International Agency for Research on Cancer
ICP	Inductively coupled plasma spectrometer
ICP-AES	Inductively coupled plasma spectrometer-atomic emission spectroscopy
IPC-MS	Inductively coupled plasma spectrometer-mass spectrometer
LBP	Lead-based paint
LBPPPA	Federal Lead-Based Paint Poisoning Prevention Act
LOD	Level of detection
MRP	Medical removal protection
NAAQS	National Ambient Air Quality Standards
NACE	National Association of Corrosion Engineers
NHANES II	National Health and Nutrition Examination Survey II
NIOSH	National Institute for Occupational Safety and Health
NIST	National Institute of Standards and Technology
NLAC	National Lead Abatement Council
NPL	National Priority List
NYCCELP	New York City Coalition to End Childhood Lead Poisoning
OSHA	Federal Occupational Safety and Health Administration
PbB	Blood lead
PEL	Permissible exposure limit
PHA	Public Housing Authority
QA	Quality assurance
QC	Quality control
RCRA	Resource Conservation and Recovery Act
SEL	Substrate equivalent lead during AXRF testing
SMSA	Standard Metropolitan Statistical Area
SOP	Standard operating procedure
SRS	Standard reevaluation schedule
SSPC	Steel Structures Painting Council
TCLP	Toxic characteristic leaching procedure
TSCA	Toxic Substances Control Act
TSP	Tri-sodium phosphate detergent
XRF	X-ray fluorescence lead paint analyzer

Appendix B

Glossary

Abatement Comprehensive process of eliminating exposure or potential exposure to lead paint and lead dust. It must include testing and measures for worker protection, containment of dust and debris, cleanup and disposal of waste, and clearance testing.

Accuracy Degree to which a measurement process determines a known amount of lead or other component in a particular reference material.

Action level Point at which something needs to be done to correct or eliminate the presence of lead.

Administrative removal Temporary removal of workers before their blood lead levels reach the point where they require medical removal.

ALC (apparent lead concentration) Average of at least 3 XRF single cycle readings on a *painted* surface.

Bias Lead reading provided by any XRF on a lead-free surface. Also called an SEL.

Biological monitoring Analysis of a person's blood and/or urine to determine level of lead contamination in the body.

Blank Nonexposed sample of the medium used for testing, such as a wipe or filter, which is analyzed like other samples to determine whether (1) samples are contaminated with lead before samples are collected (e.g., at the factory or at the testing site), (2) the samples are contaminated after sample collection (e.g., during transportation to the laboratory or in the laboratory).

CFR *Code of Federal Regulations*, the basic component of the *Federal Register* publication system. The *CFR* is a codification of the regulations of various federal agencies.

Characteristics EPA has identified four characteristics of a hazardous waste: ignitability, corrosivity, reactivity, and EP toxicity. Any solid waste that exhibits one or more of these characteristics is classified as a hazardous waste under RCRA.

CLC (corrected Lead Concentration) Difference between the ALC and the SEL.

Common area Room or area that is accessible to all tenants in a project (e.g., hallway). Generally, any area that is not kept locked.

Containment Process for protecting both workers and the environment by controlling exposures to lead dust and debris created during abatement.

Contingency plan Document setting out an organized, planned, and coordinated course of action to be followed in case of a fire or explosion or a release of hazardous waste or hazardous waste constituents from a treatment, storage, or disposal facility that could threaten human health or the environment.

Contractor Any business entity, public unit, or person performing the actual abatement for a lead abatement project.

Detection limit Minimum amount of a component that a method can reliably measure.

Direct reading XRF Analyzer that provides the operator with a display of a lead concentration calculated from the "K" X-ray intensity.

Discharge or hazardous waste discharge Accidental or intentional spilling, hazardous waste leaking, pumping, pouring, emitting, discharge, emptying, or dumping of hazardous wastes onto any land or water.

Disposal Discharge, deposit, injection, dumping, spilling, leaking, or placing of any solid waste or hazardous waste into or on any land or water so that any constituent thereof may enter the environment or be emitted into the air or discharged in any waters, including ground waters.

Disposal facility Facility or part of a facility at which hazardous waste is intentionally placed into or on any land or water, and at which waste will remain after closure.

Encapsulation Resurfacing or covering surfaces and sealing or caulking with durable materials to prevent or control chalking or flaking lead-containing substances from becoming part of house dust or accessible to children.

Engineering controls Measures implemented at the work site to contain, control, and/or otherwise reduce exposure to lead dust and debris.

EPA Identification Unique number assigned by EPA to each generator or transporter of hazardous waste and each treatment, storage, or disposal facility.

EP Toxicity Test Called the extraction procedure, a test that is designed to identify wastes likely to leach hazardous concentrations of particular toxic constituents into the ground water as a result of improper management. It is a characteristic of hazardous waste. See TCLP.

Exposure monitoring Personal air monitoring of an employee's breathing zone to determine the amount of contaminant (e.g., lead) to which he/she is exposed.

Exterior work area Outdoor porch, stairway, or other element of trim or walls on the exterior of a building.

Facility All contiguous land, structures, other appurtenances and improvements on the land, used for treating, storing, or disposing of hazardous waste. A facility may consist of several treatment, storage, or disposal operational units—that is, one or more landfills, surface impoundments, or a combination of them.

Federal Register Document published daily by the federal government that contains either proposed or final regulations (see CFR).

Final inspection Inspection by a qualified inspector, industrial hygienist, or local public health official to determine whether abatement and cleanup are complete.

Force account Term used to describe a PHA's self-performance of modernization work by the use of employees as opposed to performance by a contractor.

Generator Person who first creates a hazardous waste or first makes the waste subject to the Subtitle C regulation (e.g., imports a hazardous waste, initiates a shipment of a hazardous waste from a TSD, or mixes hazardous wastes of different DOT shipping descriptions by placing them into a single container).

Ground water Water below the land surface in a zone of saturation.

Hazardous waste As defined in RCRA, the term *hazardous waste* means a solid waste, or combination of solid wastes, that because of its quantity, concentration, or physical, chemical, or infectious characteristics may

- cause, or significantly contribute to an increase in mortality or an increase in serious irreversible, or incapacitating reversible, illness, or
- pose a substantial present or potential hazard to human health or the environment when improperly treated, stored, transported, or disposed of, or otherwise managed.

As defined in the regulations, a solid waste is hazardous if it meets one of four conditions:

1. Exhibits a characteristic of a hazardous waste (40 *CFR* Sections 261.20 through 262.24)
2. Has been listed as hazardous (40 *CFR* Sections 261.31 through 261.33)
3. Is a mixture containing a listed hazardous waste and a nonhazardous solid waste (unless the mixture is specifically excluded or no longer exhibits any of the characteristics of hazardous waste)
4. Is not excluded from regulation as a hazardous waste

HEPA (High Efficiency Particle Air) Filter capable of filtering out particles of 0.3 microns or greater from a body of air at 99.97 percent efficiency or greater.

High phosphate detergent Any enclosed device using controlled flame combustion that neither meets the criteria for classification as a boiler nor is listed as an industrial furnace.

Incinerator Any enclosed device using controlled flame combustion that neither meets the criteria for classification as a boiler nor is listed as an industrial furnace.

Indian Housing Authorities Public housing agency established by exercise of a tribe's powers of self-government independent of state law, or by operation of state law providing specifically for housing authorities for Indians.

Industrial hygienist Person certified by the American Board of Hygiene or an industrial hygienist in training, or an individual with equivalent education or experience.

Initial survey Systematic inspection of a dwelling unit by a qualified inspector, using a portable XRF analyzer, atomic absorption spectroscopy, or other approved testing techniques, to determine whether lead-based paint is present.

Interior work area Hallway, room, or group of rooms in which an abatement takes place on the inside of a building.

Landfill Disposal facility or part of a facility where hazardous waste is placed in or on land and which is not a land treatment facility, a surface impoundment, or an injection well.

Landfill liner Continuous layer of natural or man-made materials, beneath or on the sides of a surface impoundment, landfill, or landfill cell that restricts the downward or lateral escape of hazardous waste, hazardous waste constituents, or leachate.

Listed Hazardous wastes that have been placed on one of three lists developed by EPA: nonspecific source waste, specific source wastes, and commercial chemical products. These lists were developed by examining different types of waste and chemical products to see if they exhibit one of the four characteristics, meet the statutory definition of hazardous waste, are acutely toxic or acutely hazardous, or are otherwise toxic.

Logbook Notebook that accompanies each XRF analyzer, to record such information as daily performance, maintenance problems, and average reading time.

Manifest Shipping document, EPA form 8700-22, used for identifying the quantity, composition, origin, routing, and destination of hazardous waste during its transportation from the point of generation to the point of treatment, storage, or disposal.

Medical removal Temporary removal of workers due to elevated blood lead levels as defined in these guidelines.

Microgram One millionth of a gram.

Owner Person, firm, corporation, guardian, conservator, receiver, trustee, executor, or other judicial officer who, alone or jointly or severally with others, owns, holds, or controls the whole or any part of the freehold or leasehold title to any property, with or without accompanying actual possession of it, and shall include in addition to the holder of legal title any vendee in possession of it, but may not include a mortgagee or an owner of a reversionary interest under a ground rent lease.

Paint removal Strategy of abatement that entails stripping lead paint from surfaces of components.

Permit Authorization, license, or equivalent control document issued by EPA or an authorized state to implement the regulatory requirements of Subtitle C Parts 264 and 265 for TSDs.

Personal samples Air samples for sampling lead dust, collected from within the breathing zone of a worker, but outside the respirator. Samples are collected with a personal sampling pump, pulling 1–4 liters per minute of air.

Precision Degree of repeatability of a series of successive measurements.

Public Housing Agency (PHA) Any state, county, municipality, or other governmental entity or public body (or agency or instrumentality thereof) that is authorized to engage or assist in the development or operation of housing for low income families.

RCRA Resource Conservation and Recovery Act of 1976. What we commonly refer to as RCRA is an amendment to the Solid Waste Disposal Act of 1965. RCRA was amended in 1980 and most recently on November 8, 1984 by HSWA.

Replacement Strategy of abatement that entails the removal of components such as windows, doors, and trim that have lead painted surfaces and installing new components free of lead paint.

Representative sample Sample of a universe of whole (e.g., waste sample pile, lagoon, ground water, or waste stream) that can be expected to exhibit the average properties of the universe of whole.

Sample site Specific spot on a surface being tested for lead concentration through portable XRF of laboratory analysis.

SEL Substrate equivalent lead concentration is the average of at least 3 XRF single cycle readings on an unpainted surface.

Shoot To obtain the results of an XRF single reading cycle.

Site Land or water area where any facility or activity is physically located or conducted, including adjacent land used in connection with the facility or activity.

Small quantity Generator who produces less than 100 kg of hazardous waste per month (or accumulates less than 100 kg at any one time) or one who produces less than 1 kg of acutely hazardous waste per month (or accumulates less than 1 kg of acutely hazardous waste at any one time).

Sodium sulfide Chemical used to test a paint sample qualitatively for lead; typical concentrations are 6–10 percent. A positive test is characterized by a gray, black, or other dark discoloration of the paint film cross section.

Solid Waste As defined in RCRA the term *solid waste* means any garbage, refuse, sludge from a waste treatment plant, water supply treatment plant, or air pollution control facility and other discarded material, including solid, liquid, semisolid, or contained gaseous material resulting from industrial, commercial, mining, and agricultural operations, and from community activities, but does not include solid or dissolved material in domestic sewage, or solid or dissolved materials in irrigation return flows or industrial discharges that are point sources subject to permits under the Clean Water Act, or special nuclear or byproduct material as defined by the Atomic Energy Act of 1954.

Spectrum Analyzer XRF Type of XRF analyzer that provides the operator with a plot of the energy and intensity of both "K" and "L" X-rays, as well as a calculated lead concentration.

Standard Used in two ways in this manual: (1) levels established by law or regulation, such as 1.0 mg/cm^2, and (2) materials to which known quantities of lead have been applied; used to evaluate the accuracy and performance of the XRF analyzer, usually called standard reference materials.

Storage Holding of hazardous waste for a temporary period, at the end of which the hazardous waste is treated, disposed, or stored elsewhere.

Substrate Surface upon which paint or varnish has been or may be applied. Examples of substrates include wood, plaster, metal, and drywall.

Substrate effect Returning of backscattered radiation from the paint, substrate, or underlying material to the XRF analyzer. This radiation when counted as lead X-rays by an XRF contributes to SEL or bias. The inspector may have to compensate for this effect when using direct reading XRF analyzers.

TCLP Toxic characteristic leaching procedure. See EP Toxicity.

Transporter Any person engaged in the off-site transportation of hazardous waste within the United States, by air, rail, highway, or water, if such transportation requires a manifest under 40 *CFR* Part 262.

Treatment Any method, technique, or process, including neutralization, designed to change the physical, chemical, or biological character or composition of any hazardous waste so as to neutralize it, or render it nonhazardous or less hazardous, or to recover it, make it safer to transport, store or dispose of, or amenable for recovery, storage, or volume reduction.

TSD Acronym for treatment, storage, or disposal hazardous waste facility.

TSP Acronym for tri-sodium phosphate.

ULPA (Ultra low penetration air) Filter capable of filtering out particles of 0.13 microns or greater from a body of air at 99.9995 percent efficiency or greater.

XRF Analyzer Instrument that determines lead concentration in milligrams per square centimeter (mg/cm^2) using the principal of X-ray fluorescence. Two types of XRF analyzers are used, direct readers and spectrum analyzers.

Appendix C

Information Sources

GOVERNMENTAL REGULATORY AGENCIES

U.S. Department of Labor

OSHA Region I: CT, ME, MA, NH, RI, VT
33 Portland Street, 1st floor
Boston, MA 02114
(617) 565-7154

OSHA Region II: NJ, NY, PR
201 Varick Street, Room 670
New York, NY 10014
(212) 337-2378

OSHA Region III: DE, DC, MD, PA, VA, WV
Gateway Building, Suite 2100
3535 Market Street
Philadephia, PA 19104
(215) 596-1201

OSHA Region IV: AL, FL, GA, KY, MI, NC, SC, TN
1375 Peachtree Street, NE, Suite 587
Atlanta, GA 30367
(404) 347-3573

OSHA Region V: IN, IL, MI, MN, OH, WI
230 S. Dearborn Street, Room 3244
Chicago, IL 60604
(312) 353-2220

OSHA Region VI: AR, LA, NM, OK, TX
525 Griffin Street, Room 602
Dallas, TX 75202
(214) 767-4731

OSHA Region VII: IA, KS, MO, NE
911 Walnut Street, Room 406
Kansas City, MO 64106
(816) 426-5861

OSHA Region VIII: CO, MT, ND, SD, UT, WY
Federal Building, Room 1576
1961 Stout Street
Denver, CO 80294
(303) 844-3061

OSHA Region IX: AZ, CA, HI, NV, Am. Somoa, Guam, Trust Territory of Pacific Islands
71 Stevenson Street, Room 415
San Francisco, CA 94015
(415) 995-5672

OSHA Region X: AK, ID, OR, WA
Federal Office Building
909 First Avenue, Room 6003
Seattle, WA 98174
(206) 442-5930

U.S. Environmental Protection Agency (EPA)

EPA Office of Pollution Prevention and Toxics
401 M Street, SW
Washington, DC 20460
(800) 532-3394 (hotline)

EPA Region 1: CT, MA, ME, NH, RI, VT
JFK Federal Building
Boston, MA 02203
(617) 565-3424

EPA Region 2: NJ, NY, PR, VI
26 Federal Plaza
New York, NY 10278
(212) 264-2515

EPA Region 3: DE, MD, PA, VA, WV, DC
841 Chestnut Street
Philadelphia, PA 19107
(215) 597-9370

EPA Region 4: AL, FL, GA, KY, MS, NC, SC, TN
345 Courtland Street, NE
Atlanta, GA 30365
(404) 347-3004

EPA Region 5: IL, IN, MI, MN, OH, WI
230 South Dearborn Street
Chicago, IL 60604
(312) 353-2072

EPA Region 6: AR, LA, NM, OK, TX
1445 Ross Avenue
Dallas, TX 75202-2733
(214) 655-2200

EPA Region 7: IA, KS, MO, NE
726 Minnesota Avenue
Kansas City, KS 66101
(913) 236-2803

EPA Region 8: CO, MT, ND, SD, UT, WY
999 18th Street
One Denver Place, Suite 1300
Denver, CO 80202-2413
(303) 293-1692

EPA Region 9: AZ, CA, HI, NV, Am. Somoa, Guam,
Trust Territories of the Pacific
215 Fremont Street
San Francisco, CA 94105
(415) 974-8083

EPA Region 10: AK, ID, OR, WA
1200 Sixth Avenue
Seattle, WA 98101
(206) 442-1465

Department of Housing and Urban Development

Department of Housing and Urban Development (HUD)
Office of Lead-Based Paint Abatement and Poisoning Prevention
451 7th Street, SW
Room B-133
Washington, DC 20410
(202) 755-1822

Other Federal Agencies

Centers For Disease Control and Prevention (CDC)
Lead Poisoning Prevention Branch
4770 Buford Highway, NE
Building 101, Mail Stop 742
Atlanta, GA 30341
(404) 488-7330

National Institute for Occupational Safety and Health Headquarters (NIOSH)
1600 Clifton Road, NE
Atlanta, GA 30323
(404) 639-3771

National Maternal and Child Health Clearinghouse
38th and R streets, NW
Washington, DC 20057
(202) 625-8410

Public Service Campaign

On May 4, 1993, the EPA joined the National Safety Council Environmental Defense Fund and the surgeon general-designate in launching a national public service advertising campaign to heighten the public's awareness of the danger of lead poisoning in children. The EPA and the National Safety Council are sponsoring the campaign, which includes television, radio, and print ads. The ads will be used in about 50 media markets throughout the country. To obtain copies of the lead public service announcements, call the National Safety Council at (202) 833-1071. To contact the National Lead Information Center, call (800) LEAD-FYI (532-3394).

Advocacy and Professional Organizations

The Alliance to End Childhood Lead Poisoning
227 Massachusetts Avenue, NE, Suite 200
Washington, DC 20002
(202) 543-1147

The National Center for Lead-Safe Housing
205 American City Building
Columbia, MD 21004
(410) 992-0712

The National Lead Abatement Council
105 Campus Drive
Princeton, NJ 08543
(609) 520-1414

Steel Structure Painting Council
4516 Henry Street, Suite 301
Pittsburgh, PA 15213-3728
(412) 687-1113

Regional Lead Training Centers
The National University Continuing Education Association Program
One Dupont Circle
Suite 615
Washington, DC 20036
(202) 659-3130

Publications

Source of HUD publications:
HUD USER
P.O. Box 6091
Rockville, MD 20850
(800) 245-2691

For the *Federal Register*, *Code of Federal Regulations*, other federal documents:

Superintendent of Documents
U.S. Government Printing Office
Washington, DC 20402
(202) 783-3238

Depository for federal government publications, including most EPA publications:

National Technical Information Service
U.S. Department of Commerce
5285 Port Royal Road
Springfield, VA 22161
(703) 487-4650

DELEADING magazine
(trade publication and official newsletter of the National Lead Abatement Council)
P.O. Box 535
Olney, MD 20832
(301) 924-5490

Pb—SSPC's Lead Paint Bulletin
Steel Structure Painting Council
4516 Henry Street, Suite 301
Pittsburgh, PA 15213-3728
(412) 687-1113

Lead Detection & Abatement Report
(trade newsletter)
IAQ Publications, Inc.
4520 East-West Highway, Suite 610
Bethesda, MD 20814
(301) 913-0115

Lead Abatement Alert
(trade newsletter)
Lee Publishing
P.O. Box 65121
Washington, DC 20035
(301) 588-6136

Mealey's Litigation Report: Lead
Mealey Publications, Inc.
P.O. Box 446
Wayne, PA 19087-0446
(215) 688-6566

State Lead Poisoning Prevention Agencies

Alabama
Childhood Lead Poisoning Prevention Program
Alabama Department of Health
434 Monroe Street
Montgomery, AL 36130
(205) 242-5766

Alaska
Department of Health
Section of Epidemiology
P.O. Box 240249
Anchorage, AK 99524
(907) 561-4406

Arizona
Office of Risk Assessment and Investigation
3008 North Third Street
Phoenix, AZ 85012
(602) 230-5858

Arkansas
Department of Health
Division of Child and Adolescent Health
4815 West Markham, Slot 17
Little Rock, AR 72205
(501) 661-2827

California
Department of Health Services
Childhood Lead Poisoning Prevention Program
5801 Chrystie Avenue, Suite 600
Emeryville, CA 94608
(510) 450-2413

Colorado
Colorado Department of Health
4210 East 11th Avenue
Denver, CO 80220
(303)331-8373

Connecticut
Department of Health Services
Environmental Health Division
150 Washington Street
Hartford, CT 06106
(203) 566-5808

Delaware
Division of Public Health
Childhood Lead Poisoning Prevention
E.P. Bissell Hospital, Building C
3000 Newport Gap Pike
Wilmington, DE 19808
(303) 995-8696

District of Columbia
D.C. Commission of Public Health
D.C. Lead Poisoning Prevention Program
717 14th Street, NW, Suite 850
Washington, DC 20005
(202) 727-9870

Florida
State Health Office
HRS Toxicology and Hazard Assessment
1317 Winewood Boulevard
Tallahassee, FL 32399-0700
(904) 488-3385

Georgia
Childhood Lead Poisoning Prevention Program
Division of Public Health
Environmental Health Section
2 Peachtree Street, NE, 5th Floor Annex
Atlanta, GA 30303
(404) 657-6534

Hawaii
Hawaii Department of Health
1250 Punchbowl Street
Honolulu, HI 96813
(808) 733-9022

Idaho
Department of Health and Welfare
Division of Health
450 West State Street
Statehouse Mall, Boise Mall
Boise, ID 83720

Illinois
Department of Public Health
Division of Family Health
Childhood Lead Poisoning and Injury Prevention Program
535 West Jefferson
Springfield, IL 62761
(217) 782-0403

Indiana
Indiana State Board of Health
Division of Maternal and Child Health
1330 West Michigan Street
P.O. Box 1964
Indianapolis, IN 46206-1964
(317) 633-8027

Iowa
Division of Family and Community Health
Lucas State Office Building
321 12th Street
Des Moines, IA 50319-0015
(515) 242-6340

Kansas
Bureau of Environmental Health Services
109 Southwest 9th Street, Suite 501
Topeka, KS 66614
(913) 296-1690

Kentucky
Childhood Lead Poisoning Prevention Program
Division of Maternal and Child Health
275 East Main Street
Frankfort, KY 40621
(502) 564-2154

Louisiana
Office of Public Health
Genetics Division
P.O. Box 60603
New Orleans, LA 70160
(504) 568-5070

Maine
Bureau of Health
Department of Human Services
Division of Public Nursing
State House Station 11
Augusta, ME 04333
(207) 287-5690

Maryland
Maryland Department of the Environment
2500 Broening Highway
Baltimore, MD 21224
(301) 631-3861

Massachusetts
Childhood Lead Poisoning Prevention Program
305 South Street
Jamaica Plain, MA 02130
(617) 522-3700, ext. 180

Michigan
Michigan Department of Public Health
Childhood Lead Program
P.O. Box 30195
Lansing, MI 48909
(517) 335-9263

Minnesota
Minnesota Department of Health
Community and Environmental Services
925 Southeast Delaware Street
P.O. Box 59040
Minneapolis, MN 55459-0040
(612) 627-5018

Mississippi
Department of Health
Child Health Division
P.O. Box 1700
Jackson, MS 39215-1700
(601) 960-7463

Missouri
Department of Health
Bureau of Environmental Epidemiology
P.O. Box 570
Jefferson City, MO 65102
(314) 751-6102

Montana
Department of Health and Environmental Services
1400 Broadway
Helena, MT 59620-0901
(406) 444-4743

Nebraska
Department of Health
Maternal and Child Health Division
301 Centennial Mall South, Third Floor
P.O. Box 95007
Lincoln, NE 68509
(402) 471-0197

Nevada
Health, Education and Research
Maternal and Child Health Services
505 East King Street, Room 205
Carson City, NV 89701
(702) 687-4885

New Hampshire
Department of Health and Human Services
Childhood Lead Poisoning Prevention Program
Division of Public Health Services
6 Hazen Drive
Concord, NH 03301
(603) 271-4507

New Jersey
New Jersey State Department of Health
Accident Prevention and Poison Control Program
363 West State Street, CN 364
Trenton, NJ 08625
(609) 292-5666

New Mexico
Department of Health
1190 St. Francis Drive
Runnels Building, Room N1350
Sante Fe, NM 87502-6110
(505) 827-0006

New York
Health Department
Childhood Lead Poisoning Prevention Program
Empire State Plaza Building, Room 208
Albany, NY 12237
(518) 473-2762

North Carolina
Department of Environmental, Health and Natural Resources
Division of Maternal and Child Health
Childhood Lead Poisoning Prevention Program
1330 St. Mary's Street
P.O. Box 27687
Raleigh, NC 27611-7687
(919) 733-0385

North Dakota
Maternal and Child Health
Department of Health and Consolidated Laboratories
Capitol Building
Bismarck, ND 58505
(701) 224-2493

Ohio
Ohio Department of Health
246 North High Street, Sixth Floor, PSU
Columbus, OH 43266-0588
(614) 466-5332

Oklahoma
State Department of Health
Maternal and Child Health Service
1000 Northeast 10th Street
Oklahoma City, OK 73117
(405) 271-4471

Oregon
Oregon Health Division
800 NE Oregon Street, Suite 730
Portland, OR 97232
(503) 731-4025

Pennsylvania
Pennsylvania Department of Health
Division of Maternal and Child Health
Childhood Lead Poisoning Prevention Program
P.O. Box 90, Room 725
Harrisburg, PA 17108
(717) 783-8451

Rhode Island
Department of Health
Division of Family Health
302 Cannon Building
Providence, RI 02908
(401) 277-2312

South Carolina
Department of Health and Environmental Control
Division of Children's Health
2600 Bull Street
Columbia, SC 29201
(803) 737-4069

South Dakota
South Dakota Department of Health
Division of Public Health
523 East Capital
Pierre, SD 57501
(605) 773-3364

Tennessee
Department of Health
Division of Environmental Epidemiology
Cordell Hull Building, C-1 130
Nashville, TN 37247-4912
(615) 741-5683

Texas
Department of Health
Bureau of Maternal and Child Health
100 West 49th Street
Austin, TX 78756
(512) 458-7700

Utah
Bureau of Epidemiology
P.O. Box 16660
Salt Lake City, UT 84116
(801) 5380-6191

Vermont
Department of Health
Division of Environmental Health
P.O. Box 70
Burlington, VT 05402
(802) 863-7336

Virginia
Department of Health
Division of Maternal and Child Health
109 Governor Street
Richmond, VA 23219
(804) 786-7367

Washington
Department of Health
Office of Toxic Substance
Mail Stop LD-11
Olympia, WA 98504
(206) 753-2730

Wisconsin
Wisconsin Division of Health
Childhood Lead Poisoning Prevention Program
1414 E. Washington Avenue, Room 128
Madison, WI 53701
(608) 266-8154

Appendix D

Summary of Lead-based Paint Risk Assessment Process

David Jacobs, C.I.H.

1. Property owner contacts risk assessor to request services.

2. Risk assessor determines if owner is requesting a risk assessment, an inspection, or both, then reaches an agreement on costs and scope of effort. If an elevated blood lead (EBL) child is being investigated, protocols established by a local health agency may be followed. If the dwelling unit was built after 1978 (or all lead-based paint has been removed and clearance has been established), a lead-based paint risk assessment may not be warranted.

3. The property owner submits information to assessor on standard forms (or risk assessor completes forms by phone interview).

4. For single-family housing, multifamily housing consisting of different units, and multifamily housing consisting of nine or fewer similar units, conduct sampling and visual assessment in each unit. If there are 10 or more similar multifamily units, worst-case units are selected for sampling and visual assessment.

5. Perform visual assessment of building condition and paint condition, using the standard forms and protocols in this chapter.

6. Conduct dust sampling. Dust samples are typically collected at the entryway, in common spaces, and in the kitchen, child's bedroom, play area, and living room on floors, window sills, window wells, and other suspected surfaces.

7. Conduct soil sampling. Soil samples are collected from bare areas in the play area, the building foundation, and perhaps the yard (unless previous reliable neighborhood sampling shows soil levels are all below 500 ppm; if so, soil sampling may not be necessary.

8. Conduct sampling of deteriorated paint by collecting all layers of paint (not just the peeling layers) and submit to a qualified laboratory. Alternatively, deteriorated paint can be measured by portable XRF if the deteriorated paint has a large enough uniform surface with all layers present. Destructive paint chip sampling must always be done after dust sampling to avoid cross-contamination.

9. Air and water samples: If owner requests it, collect optional water samples to evaluate lead exposures that can be corrected by the owner (leaded service lines, fixtures). Water sampling is not recommended for routine risk assessments of immediate lead-based paint hazards in housing. The building owner is often unable to correct a lead-contaminated water problem, which is handled by the local water authority. Air samples are also not recommended for routine lead-based paint risk assessments.

10. Teach owner or owner's representative how to recognize lead-based paint hazards while conducting the sampling.

11. Submit samples to a qualified, certified laboratory participating in the EPA National Lead Laboratory Accreditation Program.

12. Interpret laboratory results. Prepare summary statistics (in multifamily housing with 10 or more units only).

13. Integrate lab results with visual assessment results and other maintenance and management data. Determine the presence or absence of lead-based paint hazards, as defined under statute (i.e., deteriorated paint, contaminated soil and dust).

14. Meet with the owner to discuss various lead hazard control options for specific lead hazards. Jointly determine which options are most feasible and effective for the specific situation.

15. Prepare report, recommending specific lead hazard reduction measures, including interim control and abatement options. Develop rough cost estimates of specific alternatives, including the costs of reevaluation (if applicable). The report must include a reevaluation schedule, unless all lead-based paint is removed from the dwelling.

16. Provide any certificates or other documentation required by local regulation.

Appendix E

Summary Outline of OSHA General Industry Standard (29 *CFR* 1910.1025)

David Chambers

The *OSHA General Industry Standard*, promulgated in 1978 and effective February 24, 1979, applies to all industrial settings and specifically excludes coverage of workers in construction and agriculture. This appendix is included in this textbook to provide guidance to managers of industrial facilities where lead may pose a concern both in industrial processes and in structural lead-based paint. It is a summary of major provisions of the *OSHA General Industry Standard* for lead. Readers are advised to contact OSHA for a complete copy of the standard and additional guidance or interpretation of the standard's provisions.

EXPOSURE MONITORING: 1910.1025 (d)

Generally speaking, employee exposure monitoring is conducted to determine what an employee would be exposed to if he or she was not wearing a respirator. It is the obligation of the employer to conduct full shift sampling (at least 7 continuous hours) and these samples must be truly representative of the employee's regular, daily exposure to lead.

The initial determination of employee exposures may be limited to a representative sample of the exposed employees who the employer believes are exposed to the greatest airborne concentration of lead in the workplace.

If air sample results show that employees are exposed at or above the action level of 30 μg/m^3 (30 micrograms per cubic meter) of air based on an 8-hour time weighted average (TWA), the employer shall conduct monitoring that is representative of the exposure for each employee in the workplace who is exposed to lead. If the results are below the permissible exposure limit of 50 μg/m^3 of air based on an 8-hour TWA, the employer shall continue sampling every 6 months. This schedule shall continue until at least two consecutive measurements, taken at least 7 days apart, are below the action level. However, should workplace conditions change, the employer may be obligated to reinstitute air monitoring.

If air sample results are below the action level, the employer must have a written record of this result. Air monitoring can cease unless there has been a change in workplace conditions that would warrant additional monitoring.

The employee shall be notified in writing of air monitoring results within 5 working days after the employer receives the results.

THE PERMISSIBLE EXPOSURE LIMIT (PEL): 1910.1025 (c)

The employer shall ensure that no worker is exposed to concentrations of lead greater than 50 $\mu g/m^3$ (50 micrograms per cubic meter) of air averaged over an 8-hour period.

In some circumstances, employees may be exposed to lead for greater than 8 hours. Should that be the case, OSHA provides a formula for determining the actual exposure level. The maximum permissible concentration is determined by dividing 400 $\mu g/m^3$ by the actual hours of exposure.

METHODS OF COMPLIANCE: 1910.1025 (e)

The employer must implement engineering and work practice controls to reduce employees' exposure to lead. The employer must have a written compliance program that includes:

- Description of each operation that emits lead
- Description of the means an employer will use to control exposures to lead
- Report of the technology considered in meeting the PEL
- Air monitoring data
- Detailed schedule for the implementation of the compliance program
- Work practice program that includes information regarding protective equipment, housekeeping procedures, and hygiene practices
- Administrative control schedule

An employer may use mechanical ventilation to control employee exposures. In that case, the employer must institute some sort of measurement that determines the effectiveness of ventilation, and measurements must be taken at least every 3 months. If the ventilation system includes recirculation of air, the employer must filter the air with a HEPA filter and continually monitor the system for possible failure.

RESPIRATORY PROTECTION: 1910.1025 (f)

The *OSHA General Industry Standard* places the following responsibilities on employers subject to this standard:

- Respirators must be provided at no cost to the employee
- No employer shall require a worker to wear a negative pressure respirator for more than 4.4 hours per day
- An employer must provide a respirator whenever an employee requests one. The employer must provide a powered air-purifying respirator in lieu of negative pressure respirator whenever the employee requests it and it will provide adequate protection
- All respirators must be NIOSH/MSHA approved
- The employer must have a written respiratory protection program in accordance with 29 *CFR* 1910.134 (b), (d), (e), and (f)

PROTECTIVE WORK CLOTHING AND EQUIPMENT: 1910.1025 (g)

Employers covered under this standard must provide protective clothing as follows to any employee exposed to lead above the PEL:

- Protective clothing must be provided at least weekly, and daily if the employee is exposed to greater than 200 $\mu g/m^3$ of lead as an 8-hour TWA.
- The employer is responsible for laundering protective clothing or must provide for the daily disposal of clothing. All containers of contaminated clothing will be labeled as follows: *Caution: Clothing contaminated with lead. Do not remove dust by blowing or shaking. Dispose of lead contaminated wash water in accordance with applicable local, state, or federal regulations.*

HOUSEKEEPING: 1910.1025 (h)

- All surfaces must be maintained as free as practicable of accumulations of lead
- Floors and other surfaces cannot be cleaned by compressed air and shoveling, dry or wet sweeping, and brushing can only be used if vacuuming or other equally effective methods have been tried and found not to be effective

HYGIENE FACILITIES AND PRACTICES: 1910.1025 (i)

To minimize exposure to lead, OSHA requires special attention be given to workers' personal hygiene.

- No food, drink, smoking, or applying of cosmetics in areas above the PEL
- Employees who are exposed above the PEL must be provided with a clean change room with storage facilities for street clothes and protective clothing. The storage of these articles will be such as to prevent cross-contamination
- Showers will be provided to all employees exposed to greater than the PEL and all employees must shower at the end of their shift
- Lunchrooms and lavatories will be provided to employees

MEDICAL SURVEILLANCE: 1910.1025 (j)

A medical surveillance program must be provided, at no cost to the employee, for employees who are or may be exposed to the action level for greater than 30 days per year.

- The medical surveillance program will include biological monitoring, that is, blood sampling and analysis for lead and zinc protoporphyrin levels
- Biological monitoring will take place on the following schedule:

  1. Every 6 months for employees exposed to the action level for greater than 30 days per year
  2. At least every 2 months for an employee whose last results were at or above 40 μg/100 grams of whole blood. This frequency shall continue until two consecutive results show a blood lead level below 40 μg/100g
  3. At least monthly for employees removed from the workplace due to high blood lead levels

- When an employee is removed due to an elevated blood lead level, a follow-up blood test must be conducted within two weeks after the employer receives the results of the first test
- Employees must be notified in writing within 5 working days upon receipt of the results by the employer if the blood lead level exceeds 40 μg/100g
- Medical exams must be provided to employees on the following schedule:

  1. At least annually for each employee for whom a test conducted within the past 12 months indicates a level at or above 40 μg/100g
  2. Prior to the employee being assigned to an area that is at or above the action level for lead in the air
  3. As soon as possible whenever an employee indicates possible symptoms of lead exposure; desires medical advice concerning present or past lead exposure and the effects it may have on reproduction; has demonstrated difficulty in breathing while using a respirator
  4. As medically appropriate for each employee either removed from exposure

to lead due to a risk of sustaining material impairment to health or otherwise limited pursuant to a final medical determination

The employer shall ensure that no employee, or agent of the employer, engages in prophylactic chelation at any time. Any chelation therapy must be conducted in a clinical setting under the supervision of a licensed physician.

MEDICAL REMOVAL PROTECTION: 1910.1025 (k)

Employees shall be removed from work where they are exposed to lead at or above the action level in accordance with the following schedule:

1. Any blood lead level at or above 60 μg/100g
2. Average of last three samples (or the average of all tests conducted over the previous 6 months, whichever is longer) is at or above 50 μg/100g
3. A medical determination indicates the employee may be at an increased risk of material impairment to health due to lead exposures

Employees may be returned to the former job status in accordance with the following schedule:

1. Two consecutive samples indicate a blood lead level at or below 40 μg/100g, or
2. A medical determination indicates an employee is no longer at risk of material impairment to health due to lead exposure

No employer can withhold money or benefits due an employee because of medical removal from work.

EMPLOYEE INFORMATION AND TRAINING: 1910.1025 (l)

This section of the OSHA standard mandates that employees be trained in the hazards of lead, respiratory protection, engineering controls, and other pertinent information regarding the lead compliance program if there is any chance of lead exposures at or above the action level, or for whom the possibility of skin or eye irritation exists.

SIGNS: 1910.1025 (m)

Warning signs must be placed in each work area where the PEL is exceeded. The signs shall read:

WARNING
LEAD WORK AREA
POISON
NO SMOKING OR EATING

RECORDKEEPING: 1910.1025 (n)

Exposure monitoring results shall include the date(s), number, duration, location, and results of each sample taken. A description of the sampling and analytical methods used and evidence of their accuracy is also included, as well as the type of respiratory protection the employee was using, if applicable.

For each employee sampled, the record must indicate their name, Social Security number, job classification, and a list of all other employees whose exposure the measurement is intended to represent. The employer must also note any environmental variables that could affect the measurement of employee exposures. These records are to be maintained for 40 years or for the duration of employment plus 20 years, whichever is longer.

Medical records must include the following: Name, Social Security number, and job description of employee, copy of physicians written opinion; results of any exposure monitoring of the employee or representative exposure levels provided to the physician; complaints in regards to lead exposures; medical examination results (including biological monitoring) and copies of medical and work history documentation; description of lab procedures and copies of any standards or guidelines used to interpret test results.

As with monitoring records, medical records must be maintained for 40 years of the duration of employment plus 20 years, whichever is longer.

Medical records in regards to employee removal must also be maintained. The following is required for this aspect of recordkeeping: Name and Social Security number of employee; date of removal and date of return to workplace; a brief explanation of how the removal was accomplished; and a statement indicating whether or not the removal was due to an elevated blood lead level. The medical removal records must be maintained for at least the duration of the employee's employment.

All medical records must be make available to OSHA and the employee and his or her designated representative. If a company ceases to do business, the transfer of records becomes an important issue. Section 1910.1025 states that the records must be either transferred to the successor employer or to OSHA. Once the retention period is reached for all records (i.e., 40 years have passed), the employer must notify OSHA 3 months before the disposal of these records. If so directed, the employer must give OSHA these records prior to their disposal.

OBSERVATION OF MONITORING: 1910.1025 (o)

The employer must allow an employee (or employee representative) the opportunity to observe any lead exposure monitoring conducted by the employer. The employer must provide to observers an explanation of the process, the opportunity to observe all steps taken, and the chance to record results obtained or receive copies of the results. Should the observation take place in an area that requires personal protective equipment, the employer shall provide this to the observer.

Appendix F

OSHA Fact Sheets for Lead Exposure in Construction

U.S. Department of Labor
Program Highlights

Fact Sheet No. OSHA 93-47
Lead Exposure in Construction
(#1 in a series of 6)

Worker Protection Programs

Lead has been poisoning workers for thousands of years. In the construction industry, traditionally most **over-exposures** to lead have been found in the trades, such as plumbing, welding and painting.

In building construction, lead is frequently used for roofs, cornices, tank linings, and electrical conduits. In plumbing, soft solder, used chiefly for soldering tinplate and copper pipe joints, is an alloy of lead and tin. Soft solder, in fact, has been banned for many uses in the United States. The use of lead-based paint in residential application has also been banned by the Consumer Product Safety Commission. However, since lead-based paint inhibits the rusting and corrosion of iron and steel, it is still used on bridges, railways, ships, lighthouses, and other steel structures, although substitute coatings are available.

Significant lead exposures can also arise from removing paint from surfaces previously coated with lead-based paint, such as in bridge repair, residential renovation, and demolition. With the increase in highway work, including bridge repair, residential lead abatement, and residential remodeling, the potential for exposure to lead-based paint has become more common. The trades potentially exposed to lead include iron work, demolition work, painting, lead-based paint abatement work, plumbing, heating/air-conditioning, electrical work, and carpentry/renovation/remodeling.

Operations that generate lead dust and fume include the following:

- Flame-torch cutting, welding, the use of heat guns, sanding, scraping and grinding of lead painted surfaces in repair, reconstruction, dismantling, and demolition work;
- Abrasive blasting of bridges and other structures containing lead-based paints;
- Use of torches and heat guns, and sanding, scraping, and grinding lead-based paint surfaces during remodeling or abating lead-based paint; and
- Maintaining process equipment or exhaust duct work.

The employer of construction workers is responsible for the development and implementation of a worker protection program in accordance with 29 CFR 1926.20. This program is essential in minimizing worker risk of lead exposure. Construction projects vary in their scope and potential for exposing workers to lead and other hazards. Many projects may involve limited exposure, such as the removal of paint from a few interior residential doors. Others may involve the removal, or stripping off, of substantial quantities of lead-based paints on large bridges. The employer should, as needed, consult a qualified safety and health professional* to develop and implement an effective worker protection program.

The most effective way to protect workers is to minimize exposure through the use of engineering controls and good work practices. It is OSHA policy that respirators are not to be used as the sole means of protection when engineering controls are feasible to reduce employee exposures.

This is one of a series of fact sheets highlighting U.S. Department of Labor programs. It is intended as a general description only and does not carry the force of legal opinion. This information will be made available to sensory impaired individuals upon request. Voice phone: (202) 219-8151. TDD message referral phone: 1-800-326-2577.

OSHA's new interim final standard for lead in construction limits worker exposures to 50 micrograms of lead per cubic meter of air averaged over an eight-hour workday.

At the minimum, the following elements should be included in the employer's worker protection program for employees exposed to lead:

- Hazard determination, including exposure assessment;
- Engineering and work practice controls;
- Respiratory protection;
- Protective clothing and equipment;
- Housekeeping;
- Hygiene facilities and practices;
- Medical surveillance and provisions for medical removal;
- Training;
- Signs; and
- Recordkeeping.

To implement the worker protection program properly, the employer needs to designate a competent person, i.e., one who is capable of identifying existing and predictable hazards or working conditions which are hazardous or dangerous to employees, in accordance with the general safety and health provisions of OSHA's construction standards. The competent person must have the authorization to take prompt corrective measures to eliminate such problems. Qualified medical personnel must be available to advise the employer and employees on the health effects of employee lead exposure and supeervise the medical surveillance program.

* Source for professionl safety adn health advice include insurance carries, trade organizations, state 7(c)(1) on-site consulation programs, and consultants.

GPO : 1993 O - 353-380

U.S. Department of Labor Program Highlights

Fact Sheet No. OSHA 93-48
Lead Exposure in Construction
(#2 in a series of 6)

Engineering Controls

Because lead is a cumulative and persistent toxic substance and because lead-caused health effects may result from low levels of exposure over prolonged periods of time, engineering controls and good work practices must be used where available to minimize employee exposure to lead. At a minimum, exposures must not exceed the OSHA interim final PEL of 50 micrograms per cubic meter of air (50 $\mu g/m^3$) averaged over an 8-hour period. When feasible engineering controls and work practice controls cannot reduce worker exposure to lead to at or below 50 $\mu g/m^3$, respirators must be used to supplement the continued use of engineering and work practice controls.

A competent person should review all site operations and stipulate the specific engineering controls and work practices designed to reduce worker exposure to lead. Engineering measures include local and general exhaust ventilation, process and equipment modification, material substitution, component replacement, and isolation or automation. Examples of recommended engineering controls that can be used to reduce worker exposure to lead are as follows:

- **Exhaust Ventilation**

Power tools used for the removal of lead-based paint should be equipped with dust collection shrouds or other attachments exhausted through a high-efficiency particulate air (HEPA) vacuum system. Operations such as welding, cutting/burning, heating should be provided with local exhaust ventilation. HEPA vacuums should be used during clean-up activities.

For abrasive blasting operations where full containment exists or is required, the containment structure should be designed to optimize the flow of ventilation air past the worker(s), so that the airborne concentration of lead is reduced and the visibility is increased. The affected area should be maintained under negative pressure to reduce the chances that lead dust will contaminate areas outside the enclosure. A containment structure should be equipped with dust collection and an air-cleaning device to control emissions of particulate matter to the environment.

- **Enclosure/Encapsulation**

Lead-based paint can be made inaccessible either by encapsulating it with a material that bonds to the surface, such as acrylic or epoxy coating or flexible wall coverings, or by enclosing it using systems such as gypsum wallboard, plywood panelling, and aluminum, vinyl or wood exterior siding. Floors coated with lead-based paint can be covered using vinyl tile or linoleum flooring.

The building owner, or other responsible person, should oversee the custodial and maintenance staffs and contractors with regard to all activities that involve enclosed or encapsulated lead-based paint. This will minimize potential inadvertent release of lead during maintenance, renovation, or demolition.

- **Substitution**

Zinc-containing primers covered by an epoxy intermediate coat and polyurethane topcoat are commonly used instead of lead-containing coatings.

Mobile hydraulic shears can be substituted for torch cutting under certain circumstances.

Surface preparation equipment, such as needle guns with multiple reciprocating needles completely enclosed within an adjustable shroud, can be substituted for abrasive blasting under certain operations. The shroud captures dust and debris at the cutting edge and can be equipped with a HEPA vacuum filtration system with a self-drumming feature. One such commercial unit can remove lead-based paint from flat steel and concrete surfaces, outside edges, inside corners, and pipes.

Chemical strippers used primarily on the exterior of buildings, surfaces involving carvings or molding, or intricate iron works, can be used in place of hand scraping using a heat gun. Chemical removal generates less airborne lead dust.

These strippers, however, can be hazardous and the material safety data sheets (MSDSs) for the products used must be reviewed by the employer for information on worker exposure hazards from the chemical ingredients and protective measures recommended by the manufacturer.

- **Component Replacement**

Lead-based painted building components (i.e., windows, doors, and trim) can be replaced either with new components free of lead-containing paint or with the same components after the paint has been removed off-site. Replacement is a permanent solution.

- **Process/Equipment Modification**

Brush/roller application of lead paints or other lead-containing coatings is a safer method than spraying. (Note: There is a ban on the use of lead-based paint in residential housing.) This method of application introduces little or no paint mist into the air where the mist can present a lead inhalation hazard.

Non-silica containing abrasive (e.g., steel or iron shot/grit) should be used where practical instead of sand in abrasive blasting operations. The free silica portion of the dust presents a respiratory health hazard.

Blasting techniques that are less dusty than abrasive blasting and that can be effective under some conditions include: (1) hydro- or wet-blasting (using high pressure water with or without abrasive or surrounding the blast nozzle with a ring of water), and (2) vacuum blasting where a vacuum hood for material removal is positioned around the exterior of the blasting nozzle.

Heat guns used to remove lead-based paints in residential housing units should be of the flameless electrical softener type Heat guns should have electronically controlled temperature settings to allow usage below 700°F. Heat guns should be equipped with various nozzles to cover all common applications and to limit heated work area.

When using abrasive blasting with vacuum on exterior building surfaces, care should be taken that the configuration of the heads on the blasting nozzle match the configuration of the substrate so that the vacuum is effective in containing debris.

Since HEPA vacuum cleaners can be used to clean surfaces other than just floors, operators should have attachments appropriate for use on unusual surfaces The proper use of brushes of various sizes, crevice tools and angular tools, when needed, will enhance the quality of the HEPA-vacuuming process and help reduce the amount of lead dust released into the air.

- **Isolation**

Although it is not feasible to completely enclose and ventilate some abrasive blasting operations, it is possible to isolate many operations to help reduce the potential for exposure to lead. Isolation, in this instance, consists of keeping employees not involved in the blasting operations as far away as possible from the work area. By placing the employees at a greater distance from the source of lead exposure, their exposures will be reduced.

# # #

GPO : 1993 O - 353-379

U.S. Department of Labor Program Highlights

Fact Sheet No. OSHA 93-49
Lead Exposure in Construction
(#3 in a series of 6)

Housekeeping and Personal Hygiene Practices

Lead is a cumulative and persistent toxic substance that poses a serious health risk. A rigorous housekeeping program and adherence to basic personal hygiene practices will minimize employee exposure to lead. In addition, these two elements of the worker protection program will help to prevent taking lead-contaminated dust out of the worksite and home to the workers' families, thus ensuring that the duration of lead exposure does not extend beyond the workshift and providing added protection to employees and their families.

Housekeeping

An effective housekeeping program involves at least daily removal of accumulations of lead dust and lead-containing debris. Vacuuming lead dust with high-efficiency particulate air (HEPA)-filtered equipment or wetting it with water before sweeping are effective control measures. Such cleaning operations should be conducted, whenever possible, at the end of the day, after normal operations cease. Furthermore, all persons doing the cleanup should be provided with suitable respiratory protection and personal protective clothing to prevent contact with lead.

In addition, all lead-containing debris and contaminated items accumulated for disposal should be collected and put into sealed impermeable bags or other closed impermeable containers. Bags and containers should be appropriately labeled as lead-containing waste. These measures are especially important as they minimize additional sources of exposure that engineering controls generally are not designed to control.

Personal Hygiene Practices

To minimize exposure to lead, special attention should be given to workers' personal hygiene. The employer must provide and ensure that workers use washing facilities. Clean change areas, and separate non-contaminated eating areas must also be provided. Cars should be parked where they will not be contaminated with lead. These measures will reduce the worker's period of exposure to lead and the ingestion of lead, ensure that the duration of lead exposure does not extend beyond the workshift, significantly reduce the movement of lead from the worksite, and provide added protection to employees and their families.

Change Areas: The employer must provide a clean change area equipped with storage facilities for street clothes and a separate area with facilities for the removal and storage of lead-contaminated protective work clothing and equipment. This separation is essential in preventing cross contamination of the employee's clothing.

Clean change areas are to be used for taking off street clothes, suiting up in clean working clothes (protective clothing), donning respirators prior to beginning work, and dressing in street clothes after work. No lead-contaminated items should enter this area.

Work clothing must not be worn away from the job site. Under no circumstances should lead-contaminated work clothes be laundered at home or taken from the worksite, except to be laundered professionally or properly disposed of following applicable Federal, state, and local regulations.

Showers: When there is potential for extensive contamination of the employees' skin, hair, and protective clothing, shower facilities must be provided if feasible so that exposed employees can wash lead from their skin and hair prior to leaving the worksite. Where showers are provided, employees must change out of their work clothes and shower before changing into their street clothes and leaving the worksite.

Workers who do not change into clean clothing before leaving the worksite may contaminate their homes and automobiles with lead dust. Other members of the household may then be exposed to harmful amounts of lead.

Personal Practices (eating, drinking, etc.): The employer must ensure that employees who work with lead either clean or remove their protective clothing and wash their hands and face prior to eating, drinking, smoking or applying cosmetics and that these latter practices are never permitted while in the work area or in areas subject to the accumulation of lead. HEPA vacuuming can be used to remove loose contamination from the work clothing prior to eating.

Washing Facilities: Adequate washing facilities shall be provided for employees. Such facilities shall be in near proximity to the worksite and provided with water, soap, and clean towels to enable employees to remove lead contamination from their skin.

Contaminated water from washing facilities and showers must be disposed of in accordance with applicable local, state, or federal regulations.

End-of-Day Procedures: Workers who are exposed to lead should follow these procedures upon finishing work for the day:

- Place disposable coveralls and shoe covers with the lead waste;
- Place lead-contaminated clothes, including work shoes, and personal protective equipment for laundering/cleaning (by the employer) in a closed container;
- Take a shower and wash hair; and
- Change into street clothes.

###

GPO : 1993 O - 353-381

U.S. Department of Labor Program Highlights

Fact Sheet No. OSHA 93-50
Lead Exposure in Construction
(#4 in a Series of 6)

Protective Clothing

At no cost to employees, employers must provide workers who are exposed to lead above the PEL and for whom the possibility of skin contamination or skin or eye irritation exist, clean, dry protective work clothing and equipment. Appropriate changing facilities must also be provided. Appropriate protective work clothing and equipment used on construction sites can include:

- coveralls or other full-body work clothing;
- gloves;
- vented goggles or face shields with protective spectacles or goggles; and
- welding or blasting helmets, when required.

Disposable coveralls and separate shoe covers may be used, if appropriate, to avoid the need for laundering. Non-disposable coveralls should be replaced daily. If an employee leaves the work area wearing protective clothing, the clothing should be cleaned with high-efficiency particulate air (HEPA) filter vacuum equipment to remove loose particle contamination; or as an alternative, the coveralls should be removed. Before respirators are removed, HEPA vacuuming or other suitable method, such as damp wiping, should be used to remove loose particle contamination on the respirator and at the face-mask seal. Use work garments of appropriate size, and use duct tape to reinforce their seams (e.g., underarm, crotch, and back).

Contaminated clothing that is to be cleaned, laundered or disposed of should be placed in closed containers. Containers shall be labeled with the following warning:

CAUTION: Clothing contaminated with lead. Do not remove dust by blowing or shaking. Dispose of lead-contaminated wash water in accordance with applicable local, state, or federal regulations.

Persons responsible for handling contaminated clothing should be informed of the potential hazard in writing. At no time should lead be removed from protective clothing or equipment by any means that disperses lead into the work area, such as brushing, shaking, or blowing.

At no time may workers be allowed to leave the worksite wearing lead contaminated clothing or equipment, e.g. shoes, coveralls, or head gear.

All contaminated clothing and equipment should be prevented from reaching the worker's home or vehicle. This is an essential step in reducing the movement of lead contamination from the workplace into a worker's home and provides added protection to employees and their families.

Gloves and protective clothing should be appropriate for the specific chemical exposure (e.g., solvents and caustics). Cotton gloves provide some protection against the contamination of hands and cuticles with lead dust. Workers should wear clothing that is appropriate for existing weather and temperature conditions under the protective clothing.

Heat stress: Workers wearing protective clothing can face a risk from heat stress. Additionally, heat stress may be an important concern when working in a hot environment or within containment structures. Heat stress is caused by a number of interacting factors, including: environmental conditions, type of protective clothing worn, the work activity required, and the individual characteristics of the employee.

In situations where heat stress is a concern, employers should use appropriate work/rest regimens and provide heat stress monitoring that includes measuring employee's heart rates, body temperatures, and weight loss.

A source of water or electrolytic drink should be close to the work area (in a non-contaminated eating/drinking area) so that it will be used often. Workers should wash their hands and face prior to drinking any fluid. Frequent fluid intake throughout the day will replace body fluids lost to evaporation. If such measures are used to control heat stress, protective clothing can be safely worn to provide the needed protection against lead exposure. The possibility of heat stress and its signs and symptoms should be discussed with all workers.

This is one of a series of fact sheets highlighting U.S. Department of Labor programs. It is intended as a general description only and does not carry the force of legal opinion. This information will be made available to sensory impaired individuals upon request. Voice phone: (202) 219-8151. TDD message referral phone: 1-800-326-2577.

GPO : 1993 O - 353-376

U.S. Department of Labor Program Highlights

Fact Sheet No. OSHA 93-51
Lead Exposure in Construction
(#5 in a series of 6)

Respiratory Protection

Although engineering and work practice controls are the primary means of protecting workers, source control at construction sites is often not sufficient to control exposure, and airborne lead concentrations may be high or may vary widely.

Presently, in the construction industry, respirators must often be used to supplement engineering controls and work practices whenever these controls are technologically incapable of reducing worker exposures to lead to or below 50 $\mu g/m^3$.

To provide adequate respiratory protection, respirators must be donned before entering the work area and should not be removed until the worker has left the area, or as part of a decontamination procedure. Employers must assure that the respirator issued to the employee is properly selected and properly fitted so that it exhibits minimum facepiece leakage. Respirators must be supplied by the employer at no cost to employees. Employers must perform either qualitative or quantitative fit tests for each employee wearing negative pressure respirators. Fit testing is to be performed at the time of the initial fitting and at least semiannually thereafter.

RESPIRATOR PROGRAM: When respirators are provided, the employer must establish a respiratory protection program in accordance with the OSHA standard on respirator protection, 29 CFR 1910.134.

Minimum requirements for an acceptable respirator program for lead include the following elements:

- Written standard operating procedures governing the selection and use of respirators;
- Selection of respirators on the basis of hazards to which the worker is exposed;
- Instruction and training in the proper use of respirators and their limitations;
- Regular inspection and cleaning, maintenance and disinfection; worn or deteriorated parts must be replaced, including replacement of the filter element in an air-purifying respirator whenever an increase in breathing resistance is detected.
- Storage in a convenient, clean, and sanitary location and protection against sunlight and physical damage;
- Appropriate surveillance of work area conditions and degree of worker exposure or stress (physiological or psychological) must be maintained;
- Evaluation to determine the continued effectiveness of the program;
- Physician's determination that the employee is physically able to perform the work and wear a respirator while performing the work (respirator user's medical capacity to wear and work with a respirator should be reviewed annually);
- Use of Mine Safety and Health Administration/National Institute for Occupational Safety and Health (MSHA/NIOSH) certified respirators;
- Fit testing of negative-pressure respirators;
- Breathing air used for supplied-air respirators must meet the requirements prescribed in 1910.134(d)(1); and
- Standing permission for employees to leave the work area to wash their faces and respirator face pieces whenever necessary to prevent skin irritation associated with respirator use.

RESPIRATOR SELECTION: Lead concentrations may vary substantially throughout a workshift as well as from day-to-day. The highest anticipated work concentration is to be used in the initial selection of an appropriate respirator.

Fact Sheet 92-50, "Exposure Ranges for Construction Activities," provides potential airborne exposure ranges associated with various activities commonly found in construction. These values may not be correct in every instance but can be used as guidance in making preliminary selection of respiratory types. The table, "NIOSH-recommended Respiratory Protection for Workers Exposed to Inorganic Lead," which appears here provides specific recommendations for the type of respirator to use when the actual workplace exposure reaches certain multiples of a 50 $\mu g/m^3$ permissible exposure limit (PEL). When an employer finds that exposures are lower or higher by personal air monitoring, then respirator selection can be adjusted accordingly.

In addition, if exposure monitoring or experience indicates airborne exposures to contaminants other than lead, such as solvents or polyurethane coatings, these exposures must be considered when selecting respiratory protection. A reevaluation of the respiratory protection program is required when a worker demonstrates a continued increase in blood lead levels.

ABRASIVE BLASTING AND RELATED OPERATIONS: NIOSH type CE respirators are required for use by abrasive blasting operators. Currently, NIOSH certifies ***both*** continuous flow and positive pressure respirators for abrasive blasting operations. The continuous-flow respirators are recommended by NIOSH ***only*** for airborne concentrations less than or equal to 25 times the OSHA PEL of 50 $\mu g/m^3$. Positive pressure respirators are recommended by NIOSH for airborne concentrations less than 2,000 times the OSHA PEL (50 μg/m3). Furthermore, manufacturer's instructions regarding quality of air, air pressure, and inside diameter and length of hoses must be strictly followed. Use of longer hoses or smaller inside diameter hoses than the manufacturer's specifications, or hoses with bends or kinks may restrict the flow of air to a respirator.

# # #

GPO : 1993 O - 353-383

U.S. Department of Labor Program Highlights

Fact Sheet No. OSHA 93-52
Lead Exposure in Construction
(#6 in a series of 6)

Medical Surveillance

When a construction employee is occupational exposed to lead at or above the action level of 30 $\mu g/m^3$, the employee must be provide initial medical surveillance consisting of biological monitoring in the form of blood sampling and analysis for lead and zinc protoporphyris levels. Blood lead levels are currently the best indicator of personal lead exposure. Workers potentially exposed to lead at or above the action level must be monitored for the presence of lead in the blood and the effects of lead on the blood-forming system. Full medical surveillance is to be provided to employees exposed to lead at or above the action level for more than 30 days per year. All medical examinations and consultations shall be performed by or under the direct supervision of a qualified physician and should be provided to employees at no cost, without loss of pay, and at a reasonable time and place. A qualified physician is a doctor of medicine (M.D.) or osteopathy (D.O.) familiar with the objectives and requirements of a medical surveillance program for lead exposure.

The following conditions necessitate an immediate medical consultation including, as determined by the qualified physician, a physical examination and a venous blood sample for lead analysis (biological monitoring):

- whenever a worker develops signs or symptoms associated with lead toxicity; and
- before a worker restarts work following medical removal.

1. Biological Monitoring

The purpose of biological monitoring is to identify workers with elevated blood lead levels. The data from biological monitoring is objective evidence of a worker's body burden from lead exposure and this data can be used to follow changes in worker exposure.

Blood lead and zinc protoporphyrin or free erythrocyte protoporphyrin (FEP) should be monitored for those workers exposed to lead. The qualified physician should determine the required frequency for biological monitoring on an individual basis. In general, workers in high-risk occupations should be monitored as often as needed to prevent adverse health effects.

Analysis of blood samples should be conducted by a laboratory currently approved by OSHA. Employers should contact their local OSHA area office for a current list of approved labs.

2. Reproductive Hazard Issues

Lead is toxic to both male and female reproductive systems. Workers who are actively seeking to have a child or who are pregnant should contact qualified medical personnel to arrange for a job evaluation and medical follow-up. Employers who have been contacted by employees with concerns about reproductive issues should refer them to qualified medical personnel.

3. Written Medical Opinion

Employers must obtain a written signed opinion from the examining physician for each medical examination performed for each employee. This opinion should contain the results of the medical examination as they relate to occupational exposure to lead and must include:

- whether the employee has any detected medical condition which would place his/her health at increased risk from lead exposure;
- any special protective measures or limitations on worker's exposure to lead;
- any limitation on respirator use;
- results of blood lead determination; and
- a statement that the employee has been informed by the physician of the results of the consultation or medical examination and any medical condition that may require further examination or treatment.

Findings of lab results or diagnoses unrelated to the workers' exposure to lead must not be communicated to employers or included in a written opinion.

Employees should be advised by each physician of any medical condition, occupational or non-occupational, which necessitates further medical evaluation or treatment. The

This is one of a series of fact sheets highlighting U.S. Department of Labor programs. It is intended as a general description only and does not carry the force of legal opinion. This information will be made available to sensory impaired individuals upon request. Voice phone: (202) 219-8151. TDD message referral phone: 1-800-326-2577.

employer should furnish the employee with a copy of the written medical opinion.

4. Chelation

The use of chelating drugs as a prophylactic measure (i.e., to prevent a detectable rise in blood lead) is an unacceptable medical practice. Chelation may be used by a qualified physician only for diagnostic or therapeutic reasons (that is, to diagnose or treat the signs and symptoms of severe lead toxicity).

5. Medical Removal

Medical removal will protect worker health both by stopping further occupational exposure and by enabling the worker to excrete the absorbed lead naturally. With good engineering, work practices, personal hygiene, and respiratory protection practices in place, very few employees should reach the medical removal trigger level specified in the OSHA standard.

OSHA's interim final standard for lead in construction uses a medical removal trigger level of 50 μg/dl. However, some authorities believe that medical removal should take place at lower levels.

Presently, 15 states require laboratories and health care providers to report cases of elevated blood lead concentrations to their state Health Departments. A list of the states that require such reporting, contact person, and the concentration that requires reporting for each state can be found in the **NIOSH Alert:** ***Preventing Lead Poisoning In Construction Workers.***

When employees are removed, or otherwise limited, they must be placed in jobs that will not result in exposure to lead at or above the action level of 30 $\mu g/m^3$. The employer may return the employer to his or her former job status when a qualified physician's medical determination is that the employee is no longer at risk from exposure to lead or when the employee's blood lead level drops below 40 μg/dl.

In the case of medical removal, records must include the following information:

- the name and social security number of the worker;
- the date of each occasion that the worker was removed from current exposure to lead;
- the date on which the worker was returned to his or her former job status;
- a brief explanation of how each removal was or is being accomplished; and
- a statement indicating whether or not the reason for the removal was an elevated blood lead level.

The employer must maintain this record for at least the duration of any worker's employment.

6. Recordkeeping

The employer must maintain any employee exposure and medical records to document ongoing employee exposure, medical monitoring and medical removal of workers. This data provides a base to properly evaluate the employee's health.

Employers must properly record cases on their OSHA form 200 when the worker:

- has a blood lead level that exceeds 50 mg/dl;
- has symptoms of lead poisoning, such as colic, nerve damage, renal damage, anemia, or gum problems; or
- receives medical treatment to lower blood lead levels or for lead poisoning.

In addition, employees or former employees, their designated representatives, and OSHA must be provided access to exposure and medical records in accordance with 29 CFR 1910.20.

###

GPO : 1993 O - 353-375

Appendix G

National Lead Abatement Council Mission Statement

NLAC is committed to conducting various education, research, legislative, and business activities related to lead-based paint. These activities will be conducted in an ethical, scientific, and professional manner. NLAC will:

Provide a scientific forum for the exchange of informaiton on state-of-the-art lead-based paint detection and abatement methods through various symposia, meetings, and publications;

Initiate activities to properly train and accredit lead-based paint detection and abatement contractors;

Promote legislative, legal, and regulatory efforts to further enhance the professional nature of lead-based paint detection and abatement, and work to assure that such activities do not create new hazards;

Promote efforts to provide sufficient funding for lead-based paint detection and abatement work in private homes, public buildings, and schools and to assure that such activities do not create new hazards;

Petition the Occupational Safety and Health Administration to expand the scope of the general industry lead standard to include construction activities, which will help ensure that workers who are exposed to lead during abatement activities are entitled to the same levels of protection accorded to those employed in industry;

Develop a multiyear long-range plan to permit a rational comprehensive approach to solving the complex lead-based paint problem.

Index